David Hillerkuss

Single-Laser Multi-Terabit/s Systems

Karlsruhe Series in Photonics & Communications, Vol. 9
Edited by Profs. J. Leuthold, W. Freude and C. Koos

Karlsruhe Institute of Technology (KIT)
Institute of Photonics and Quantum Electronics (IPQ)
Germany

Single-Laser Multi-Terabit/s Systems

by
David Hillerkuss

Dissertation, Karlsruher Institut für Technologie (KIT)
Fakultät für Elektrotechnik und Informationstechnik, 2012

Impressum

Karlsruher Institut für Technologie (KIT)
KIT Scientific Publishing
Straße am Forum 2
D-76131 Karlsruhe
www.ksp.kit.edu

KIT – Universität des Landes Baden-Württemberg und
nationales Forschungszentrum in der Helmholtz-Gemeinschaft

KIT Scientific Publishing 2013
Print on Demand

ISSN 1865-1100
ISBN 978-3-86644-991-6

Single–Laser Multi–Terabit/s Systems

Zur Erlangung des akademischen Grades eines

DOKTOR-INGENIEURS

der Fakultät für Elektrotechnik und Informationstechnik
des Karlsruher Instituts für Technologie

genehmigte

DISSERTATION

von

Dipl.-Ing. David Hillerkuß

aus

Karlsruhe

Tag der mündlichen Prüfung: 12.07.2012
Hauptreferent: Prof. Dr. sc. nat. Jürg Leuthold
Korreferenten: Prof. Dr.-Ing. Dr. h. c. Wolfgang Freude
Prof. Dr.-Ing. Michael Hübner

To my beloved family

"In the middle of every difficulty lies opportunity."

„In der Mitte der Schwierigkeit liegt die Möglichkeit.“
– **Albert Einstein**

Table of Contents

Table of Figures

Abstract

Diese Arbeit befasst sich mit der Entwicklung von Kommunikationssystemen, in welchen mittels eines einzigen Lasers Datenraten von mehr als 10 Tbit/s übertragen werden können. Innerhalb der letzten 20 Jahre ist das Verkehrsaufkommen in den Kernnetzen des Internets um mehr als das 2000-fache gestiegen. Auch gehen aktuelle Prognosen für die nächsten Jahre von einem kontinuierlichen Wachstum um 32 % pro Jahr aus. Es ist daher unumgänglich, diesen Zuwachs an Datenrate durch einen Ausbau der Kommunikationsnetze aufzufangen. Dabei ist darauf zu achten, dass dies unter möglichst geringen Kosten und energieeffizient erfolgt. Bisher konnten höhere Geschwindigkeiten durch Nutzung einer größeren Bandbreite in der Glasfaser erzielt werden. Während die Glasfaser eine nutzbare Bandbreite von 60 THz bietet, ist die Bandbreite der erbium-dotierten Glasfaserverstärker auf ca. 11 THz begrenzt. Diese Verstärkerbandbreite wird von heutigen Kommunikationssystemen bereits ausgenutzt. Während man für kurze Strecken über den Einsatz von optischen Halbleiterverstärkern nachdenken kann, sind solche Verstärker für Kommunikationsverbindungen höherer Reichweite aufgrund des größeren Rauschens und oft starken Kanalübersprechens nicht denkbar. Wenn auf den Einsatz von weiteren oder neuen Glasfasern verzichtet werden soll, so bleibt eigentlich nur noch der Ansatz, möglichst bandbreiteneffizient zu arbeiten um die nutzbare Bandbreite besser auszunutzen. Dabei gibt es zwei Techniken, mir denen die spektrale Effizienz erhöht werden kann: Die Nutzung von Modulationsformaten höherer Ordnung ermöglicht auf einem einzelnen Kanal die Übertragung höherer Datenraten bei gleichbleibender Bandbreite, während raffiniertere Multiplexverfahren die ungenutzte Bandbreite zwischen den einzelnen Kanälen reduzieren. Im Rahmen dieser Arbeit werden nach einer theoretischen Einführung zunächst die für den Einsatz dieser Techniken notwendigen Komponenten vorgestellt. Systemexperimente zeigen, dass diese Komponenten geeignet sind um Datenraten bis zu 32,5 Tbit/s mit einem einzigen Laser zu übertragen.

In der Einführung werden zunächst die verschiedenen in dieser Arbeit eingesetzten Modulationsverfahren und die dazu benötigten Sender und Empfänger erklärt. Dabei werden die Grundlagen für das Verständnis der Quadraturamplitudenmodulation (QAM) und des optischen Multiformattransmitters gelegt. In einem zweiten Teil werden die in dieser Arbeit verwendeten Multiplexverfahren eingeführt. Im Gegensatz zu den traditionell verwendeten Multiplexverfahren der optischen Nachrichtentechnik, Zeitmultiplex und Wellenlängenmultiplex, erreichen orthogonale Frequenzmultiplexverfahren (OFDM) und Nyquist-Wellenlängenmultiplexverfahren (Nyquist-WDM) die theoretische Effizienzgrenze, die durch die Nyquist-Bandbreite der Signale vorgegeben ist. Der letzte Teil der Einführung behandelt einige ausgewählte Komponenten, welche bei Einsatz digitaler Signalverarbeitung verwendet werden.

Technologien für Terabit/s Kommunikationssysteme

Durch eine Kombination von Phasen- und Amplitudenmodulation wird bei Modulationsformate höherer Ordung eine Steigerung der Datenraten bei konstanter Signalbandbreite erreicht. Um solche Modulationsformate einsetzen zu können, wird ein neuartiger optischer Transmitter benötigt. Innerhalb dieser Arbeit wurde ein software-definierter, optischer Multiformattransmitter entwickelt. Dieser Transmitter erzeugt abhängig von seiner Programmierung verschiedene Modulationsformate höherer Ordnung. Diese Flexibilität wurde durch den Einsatz von programmierbarer Hardware, so genannter Field-programmable-gate-arrays (FPGA), und schneller Digital-Analog-Wandlern (DAC) erreicht. Dabei wird das erzeugte Modulationsformat in Echtzeit in digitaler Signalprozessierung erzeugt. Der Transmitter generiert verschiedene Modulationsformate, von einem einfachen An-Ausschalten des optischen Trägers bei On-Off-Keying (OOK) über binäre Phasenmodulation (BPSK) und Quadraturphasenmodulation (QPSK) bis hin zur Quadraturamplitudenmodulation mit 16 möglichen Zuständen (16QAM). Dabei wurden Symbolraten von bis zu 30 GBd erreicht. Bei Verwendung von 16QAM als Modulationsformat und unter Verwendung von Polarisationsmultiplex wurde eine Datenrate von 160 Gbit/s gezeigt. Für die Verwendung in einem Nyquist-WDM-System wurde der Transmitter noch um Fähigkeiten zur Pulsformung erweitert. Die notwendige Überabtastung reduziert allerdings die maximal mögliche Datenrate. Auf Basis dieses Transmitters wurden in hier nicht erörterten Arbeiten Datenraten von bis zu 336 Gbit/s erzielt und der erste OFDM-Transmitter mit Signalverarbeitung in Echtzeit und einer Datenrate über 100 Gbit/s vorgestellt.

Optische Träger mit einem präzisen Frequenzabstand sind für den Einsatz von OFDM und Nyquist-WDM essentiell. Für die Orthogonalität von OFDM-Unterträgern ist ein fester Frequenzabstand vorgeschrieben. Andernfalls kommt es direkt zu Übersprechen. Da bei Nyquist-WDM die Bandbreite optimal ausgenutzt werden soll, bleibt kein Platz für spektrale Schutzbänder, welche eine Frequenzdrift der einzelnen Träger zueinander erlauben würden. Um eine relative Frequenzdrift von vornherein ausschließen zu können, werden oft Frequenzkammgeneratoren eingesetzt. Diese Kammquellen haben den entscheidenden Vorteil, dass der Frequenzabstand der erzeugten Träger durch den Aufbau der Generatoren fest vorgegeben ist. Falls nun Drifteffekte auftreten, so verschieben sich alle Träger gemeinsam, und ihr Frequenzabstand bleibt konstant. In dieser Arbeit wird ein Frequenzkammgenerator vorgestellt, der in einer spektralen Bandbreite von 4,05 THz eine Anzahl von 325 Trägern gleicher Leistung erzeugen kann. Diese Träger sind qualitativ so hochwertig, dass sie für 16QAM Übertragung geeignet sind. Übertragungsreichweiten von mehr als 200 km sind möglich. Dieser Frequenzkamm war voraussetzung für die Experimente mit den enormen Datenraten, welche am Ende der Arbeit vorgestellt werden.

Um OFDM-Signale mit Datenraten von 10 Tbit/s oder mehr zu prozessieren, wird optische Signalverarbeitung benötigt. Dies kann mit Elektronik nicht realisiert werden. Die Kernkomponente von OFDM-Sendern und Empfängern ist die schnelle Fourier-Transformation (FFT). In dieser Arbeit wird ein Schema zum Aufbau der optischen FFT vorgestellt und angewendet, wobei die Komplexität gegenüber einer direkten Implementierung der FFT deutlich reduziert wird. Die Anzahl der erforderlichen Elemente zum Aufbau der FFT steigt hier nur linear mit der Anzahl der zu verarbeitenden Stützstellen N der FFT an, nicht mit $N \log_2 N$, wie im Fall einer direkten Implementierung des FFT-Algorithmus. In einem ersten Experiment bei 400 Gbit/s wurde die Funktionalität des Konzeptes nachgewiesen und damit der Grundstein für die im Folgenden beschriebenen Systemexperimente gelegt.

Terabit/s-Datenübertragungsexperimente

Nachdem mit den entwickelten Technologien und Komponenten alle Bausteine für den Aufbau der Kommunikationssysteme verfügbar standen, wurden zwei optische OFDM-Experimente und ein Nyquist-WDM-Experiment durchgeführt.

Im ersten OFDM-Experiment wurde erstmals eine Datenrate von mehr als 10 Tbit/s mit nur einem einzigen Laser übertragen. Dabei wurden auf 75 Trägern mit einem Abstand von 25 GHz QPSK und alternativ 16QAM bei 18 GBd übertragen. Datenraten von 5,4 und 10,8 Tbit/s wurden erreicht. Die höchste zuvor erzielte Datenrate in einem OFDM-Experiment war 1,5 Tbit/s. Diese Steigerung gegenüber dem damaligen Stand der Technik wurde durch den Einsatz von 16QAM und durch die für diese Arbeit entwickelte neuartige Kammquelle ermöglicht. In diesem Experiment wurde keine Übertragungsstrecke eingesetzt. Am Empfänger wurden die OFDM-Unterträger mit einer optischen 8-Punkt-FFT separiert und im Anschluss kohärent empfangen. Der FFT basierte Empfänger war zwei weiteren untersuchten Empfängerkonzepten klar überlegen.

Zur Untersuchung der Übertragungseigenschaften wurde ein zweites OFDM-Experiment bei 26 Tbit/s durchgeführt. Dabei wurde durch Einsatz einer verbesserten Kammquelle die Datenrate nochmals deutlich gesteigert. Das Signal mit 325 Trägern in einem Abstand von 12,5 GHz wurde mit einer Symbolrate von 10 GBd gesendet. Als Modulationsformat kam erneut 16QAM zum Einsatz. Das Signal wurde über eine Übertragungsstrecke bestehend aus 50 km Standard-Einmodenfaser (SMF) und einer anschließenden dispersionskompensierenden Faser übertragen. Das Signal wurde im Anschluss mittels der optischen FFT empfangen. Auch wurde die Toleranz eines solchen OFDM-Empfängers gegenüber einer restlichen, nicht kompensierten Dispersion untersucht.

In einem dritten Experiment wurde Nyquist-WDM als Alternative zu OFDM untersucht. In diesem Experiment wurde der Aufbau des 26 Tbit/s-Experimentes angepasst, um stattdessen ein 32.5 Tbit/s-Nyquist-WDM Signal zu übertragen. Als Übertragungsstrecke kamen in diesem Experiment bis zu 227 km SMF zum Einsatz. Allerdings wurde im Gegensatz zu dem OFDM-Experiment keine optische Dispersionskompensation benötigt. Dabei stellte sich heraus, dass Nyquist-WDM eine größere Übertragungsdistanz ermöglichte.

Abschließend lässt sich sagen, dass sich sowohl optisches OFDM als auch Nyquist-WDM für bandbreiteneffiziente Kommunikationssysteme der nächsten Generation eignen, wobei Nyquist-WDM bei Betrachtung der Übertragungsdistanz das größere Potential zu haben scheint.

Preface

Super-channels for multi-Terabit/s transmission are seen as the future in optical communication systems. In this work, single-laser multi-Terabit/s transmission systems are investigated to cope with the traffic growth expected for the next years. From 1990 to 2010, the traffic in optical core networks has increased by a factor of 2000. For the next years, current estimates predict a growth in internet traffic by approximately 32 % per year. All internet traffic that is transmitted over significant distances is transported by optical communication systems. It is therefore vital, that these systems are able to handle the expected data rates at reasonable cost and power consumption. Up to this date, additional traffic was transmitted through existing fiber plants by using additional parts of the available spectrum. While optical fibers offer several tens of THz of usable bandwidth, erbium doped fiber amplifiers (EDFA) offer only approximately 11 THz of amplification bandwidth. This bandwidth can be fully used with state-of-the-art communication systems. While semiconductor optical amplifiers (SOA) would be an option for short reach applications, they cannot be used for large transmission distances due to the channel crosstalk and the large amount of noise introduced by these amplifiers. One option would be to use a special multiplexing by using additional optical fibers. However, this increases cost and energy consumption significantly. So there are only two options left to increase the transmission capacity. The first option is to make use of advanced modulation formats like quadrature amplitude modulation (QAM) to transmit more data within the same signal bandwidth. The second option is to avoid the guard bands used in standard transmission systems based on wavelength division multiplexing (WDM) through the use of advanced optical multiplexing schemes. Here, orthogonal frequency division multiplexing (OFDM) and Nyquist wavelength division multiplexing (Nyquist WDM) are investigated.

This work commences with an introduction into the topic. After presenting the theoretical background, the different subsystems or components are presented that are necessary for the following Terabit/s system demonstrations with advanced multiplexing schemes.

Chapter 1 is an introduction into the topic. After discussing traffic increase and bandwidth constraints in optical communication systems a first introduction into modulation formats and multiplexing schemes follows.

Chapter 2 covers the theoretical background on modulation formats and multiplexing schemes. The theoretical background in real-valued modulation, complex modulation, and modulation formats applied in this work are introduced. Different transmitter and receiver schemes for these formats are presented. The theoretical background of OFDM, Nyquist WDM and the relation between these two formats is given. Some details on digital signal processing are also provided.

Technologies for Terabit/s Communication Systems

Three key technologies were developed within this work that enabled the Terabit/s transmission systems, the multi-format transmitter, an optical comb source, and an optical fast Fourier transform.

Chapter 3 covers the flexible optical multi-format transmitter [1]. To be able to use different advanced modulation formats, a novel and flexible transmitter was needed. To achieve full flexibility, this transmitter is based on programmable hardware, so-called field programmable gate arrays (FPGA), and high speed digital-to-analog converters. The output signal of the transmitter is computed in real-time by the FPGA, and the programming of the FPGA defines the modulation format generated. Modulation formats from on-off-keying (OOK), over binary and quadrature phase shift keying (BPSK and QPSK), to 16-ary quadrature amplitude modulation (16QAM) are generated by the transmitter. The maximum achieved symbol rate of the transmitter is 30 GBd. For 16QAM, a total aggregate data rate with polarization multiplexing of 160 Gbit/s is presented. In follow-up work, this transmitter was extended to generate 336 Gbit/s dual-polarization 64QAM signals, 101 Gbit/s single polarization 16QAM-OFDM, and Nyquist pulse-shaped signals with dual polarization 64QAM at up to 150 Gbit/s in real-time. This Nyquist pulse-shaping transmitter was then also used in the Nyquist WDM experiments.

Chapter 4 introduces the optical comb generator. High quality optical frequency combs are crucial for advanced multiplexing schemes in next generation optical communication systems. Even though extremely broad frequency combs have been used for measurement applications, the quality requirements for high speed communications are different. It is challenging to generate extremely broad frequency combs while maintaining a high carrier quality that is suitable for communication signals. For this work, frequency combs with highest bandwidths for communication experiments have been generated using spectral broadening in a highly nonlinear fiber However, the achievable bandwidth in this scheme is limited due to the minima that appear in spectra broadenend by self phase modulation [2]. To remove these minima, spectral composing was used. In this process, the stable and high-power parts of one or more broadened spectra and the original spectrum are combined to form a stable output spectrum. The generated frequency comb comprises 325 optical carriers with a spacing of 12.5 GHz. This corresponds to a total bandwidth of 4.05 THz. The quality of all carriers is suitable to transmit 16QAM data over distances beyond 227 km. In transmission experiments, aggregate data rates of up to 32.5 Tbit/s are achieved.

Chapter 0 contains a detailed technique, simulations, and an experimental verification of the optical fast Fourier transform (FFT). The fast Fourier transform is a universal mathematical tool for almost any technical field. It is the key component for next generation signal processing such as required for OFDM in optical communications. Here, a practical implementation of the optical FFT is introduced. Feasibility, tolerances towards further simplifications, practical implementations, and an experimental verification of the scheme demonstrate the potential for a next generation OFDM system. A proof-of-principle experiment supports the applicability to optical communication systems. Here, a single 400 Gbit/s OFDM channel is demultiplexed by an optical FFT. The method is based on passive optical components only and thus basically provides processing without any power consumption. The scheme is in support of a new paradigm, where all-optical and electronic processing synergistically interact, providing their respective strengths. All-optical methods allow processing at highest speed with little power consumption, and electronic processing performs the fine granular processing at medium to low bit rates.

Terabit/s system experiments

Chapter 6 covers three Terabit/s system experiments, two with all-optical OFDM and one with Nyquist WDM.

In a first all-optical OFDM experiment, a total data rate of 5.4 and 10.8 Tbit/s is transmitted using a single laser. This is the first experiment in which a data rate larger than 10 Tbit/s has been encoded onto a single laser. The highest data rate reported for an OFDM signal before this experiment was 1.5 Tbit/s. The novel optical comb source and 16QAM as modulation format made this significant increase in data rate possible. The OFDM signal consisted of 75 subcarriers with a spacing of 25 GHz. The symbol rate was 18 GBd. The all-optical OFDM receiver comprised an optical eight-point FFT, a bandpass filter, and subsequent coherent detection. In further tests, the performance of the FFT was compared to two other receiver schemes. In this comparison, the FFT-based receiver exhibited superior performance.

To investigate transmission performance of such an all-optical OFDM signal, a follow-up experiment with a 26 Tbit/s all-optical OFDM signal was performed. Here, the total data rate was increased further through the use of an improved optical comb source. This comb source generated 325 optical carriers with a spacing of 12.5 GHz, which were modulated with 16QAM data at 10 GBd. The signal was transmitted over a distance of 50 km of standard single mode fiber (SSMF – SMF-28) with optical chromatic dispersion compensation. In the receiver, an optical FFT was used to separate the subcarriers before coherent reception.

In the third experiment, single-laser 32.5 Tbit/s 16QAM Nyquist WDM transmission was investigated. The total transmission distance was of 227 km of SMF-28 without optical dispersion compensation. Again, the 325-carrier optical comb source is used, which is now encoded with dual-polarization 16QAM data using sinc-shaped Nyquist pulses. The symbol rate is chosen equal to the carriers spacing of 12.5 GHz. A net spectral efficiency of 6.4 bit/s/Hz is achieved. The multi-format transmitter generates the electric drive-signals for the electro-optic modulator in real-time.

Chapter 7 contains a summary of this work and an outlook into future research on advanced multiplexing and modulation formats for optical communication systems.

Achievements of the Present Work

In this thesis, single laser Tbit/s communication systems have been investigated and experimentally demonstrated. Several key components of the Tbit/s communication systems had to be developed to achieve the demonstrated speeds.

In the following, a brief summary of the main achievements is provided.

Technologies for Single–Laser Tbit/s Communication Systems

Flexible Optical Multi-Format Transmitter: Flexible transmitters can easily adapt to changing bandwidth demands and changing channel quality. The signals are generated through real-time digital signal processing (DSP) in field programmable gate arrays (FPGA) and converted to the analog domain with high speed digital-to-analog converters (DAC). The transmitter modulates an optical carrier with on-off-keying (OOK), binary phase shift keying (BPSK), QPSK, and 16QAM. The transmitter operates at a symbol rate of up to 30 GBd. This transmitter was the first optical multi-format transmitter operating at these speeds [1]. The flexible design of the transmitter allows for an easy extension in follow up work. Up to now, several more modulation schemes like 8PSK, 32QAM and 64QAM [3], orthogonal frequency division multiplexing (OFDM) at more than 100 Gbit/s [4], and pulse shaping [5, 6] have been implemented.

Optical Comb Generation: Spectral broadening of a pulse train from a mode-locked laser (MLL) in a highly nonlinear fiber (HNLF) is used to generate a large number of optical carriers. In a first experiment, a number of 75 optical carriers with a spacing of 25 GHz were generated [7]. In a second experiment, this scheme was improved by combining the original MLL spectrum and the broadened spectrum to compose a comb spectrum with 325 optical carriers and a spacing of 12.5 GHz [8]. This corresponds to an absolute bandwidth larger than 4 THz and is currently the only comb source that covers such a spectral width with a carrier quality suitable for optical transmission systems.

Optical Signal Processing – The Optical Fast Fourier Transform (FFT): The FFT algorithm can directly be implemented in an optical ciruit. The (inverse) fast Fourier transform ((I)FFT) has a reduced complexity when compared to the (inverse) discrete Fourier transform. The complexity of the (I)FFT grows proportional to $N \log_2(N)$. The symbol N represents the number time or frequency samples processed by a given FFT or IFFT respectively. This is also known as the (I)FFT size. When combining the optical FFT with a parallel to serial conversion, it can be simplified further by a transformation of the corresponding optical circuit. The complexity of the simplified implementation of the optical (I)FFT then grows only proportional to N. In addition, instead of having a number of $\log_2(N)$ cases where N optical phases have to be controlled with respect to each other, there are now only N-1 cases, where 2 relative optical phases have to be controlled. For the case where only one output of the (I)FFT is of interest, the implementation complexity is reduced even more. When combining the optical FFT with optical filters, the complexity can be reduced further.

This work was awarded with the first prize of the outstanding student paper competition at the optical fiber communication conference 2010 [9, 10].

Terabit/s Transmission System Demonstrations

Having developed the required components, several world record experiments were performed:

First Single-Laser 5.4 and 10.8 Tbit/s All-Optical OFDM Transmitter and Receiver: An optical OFDM signal with 75 subcarriers was generated and received. The subcarriers were encoded with QPSK or 16QAM at 18 GBd in multi-format transmitters. While the total data rate for QPSK was 5.4 Tbit/s, 10.8 Tbit/s were transmitted in the 16QAM experiment. An optical FFT with a size of $N = 8$ in combination with optical filters was used to separate the subcarriers. When compared to narrow filtering and a coherent receiver only, the optical FFT-based receiver clearly outperformed the two other receiver schemes tested in the experiment. Net spectral efficiencies of 2.7 bit/s/Hz for QPSK modulation and 4.6 bit/s/Hz for 16QAM modulation were achieved. This was the first experiment to show data rates beyond 10 Tbit/s encoded on a single laser [7].

First Single-Laser 26.0 Tbit/s All-Optical OFDM Transmission Experiment: The work on the 10.8 Tbit/s experiment was extended in a subsequent experiment. Here, 325 optical carriers with a spacing of 12.5 GHz were generated in the improved optical comb source. Multi-format transmitters modulated the carriers with 16QAM at 10 GBd. The signal was transmitted over 50 km of standard single mode fiber with subsequent optical dispersion compensation. The signal was received and analyzed in an all-optical OFDM receiver with an optical FFT of the size $N = 2$. The net spectral efficiency was increased to 5 bit/s/Hz. The achieved data rate in this experiment is still the highest data rate in an OFDM experiment using a single laser [8].

First Single-Laser 32.5 Tbit/s Nyquist WDM Transmission Experiment: Using Nyquist WDM, data rate and transmission reach were increased further. As in the 26 Tbit/s experiment, 325 carriers with a spacing of 12.5 GHz were used. Here however, the carriers were encoded with sinc-shaped Nyquist pulses that carry 16QAM data at 12.5 GBd. The multi-format transmitter generated sinc-shaped Nyquist pulses in real-time using a finite impulse response filters. The signal with 32.5 Tbit/s was transmitted over a distance of 227 km, which is more than 4 times longer than in the 26 Tbit/s experiment. The spectral efficiency was increased to 6.4 bit/s/Hz. This experiment was the first Nyquist WDM experiment with real-time DSP and it is the highest data rate encoded on a single laser so far [12-15].

1 Introduction

In our modern information society, we strongly depend on communication and the access to information – may it be at work or in our private life. To communicate, we use technologies like telephone, e-mail, instant messaging, video conferencing, and social networking platforms. Also, the internet provides access to a vast amount of information like e.g. news, scientific articles, books, online video and music. In the past decade it has become the most important source for knowledge.

This has led to a tremendous increase in internet traffic. While the global internet traffic in 1990 was only 1 Terabyte/month [16], it has grown by a factor of 2000 until 2010 [17]. Currently, traffic grows by approximately 32 % per year and is expected to grow from 20 Exabyte/month in 2010 to more than 80 Exabyte/month in 2015 [17]. This corresponds to a total IP traffic of approximately 1 Zetabyte or 10^{21} byte in 2015. From the graph displayed in Fig. 1.1, it is clear that the coming traffic growth is mainly caused by video services for consumers, while the contribution from internet browsing, e-mail, and file sharing is smaller. It is expected that new services like cloud computing, 3D high-definition TV and virtual reality applications will contribute significantly to the increase in data rates [8].

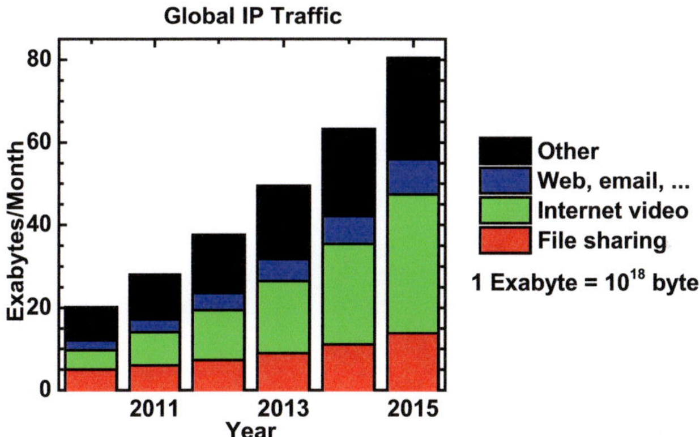

Fig. 1.1: Global IP traffic according to the Cisco Visual Networking Index [17] grows by approximately 32 % per year. Video services are expected to contribute most to the growth of internet traffic until 2015.

Both, the communication systems and the internet itself, rely heavily on optical communication systems for the transmission of data. Optical transmission links are now the most commonly used links for distances larger than hundred meters.

The reason for this is the tremendous bandwidth and the low loss of glass fibers (see Fig. 1.2a) when compared to copper cables. While the highest quality coaxial copper cables that are used as patch cords in laboratory applications have bandwidths of up to 110 GHz and losses in the range of several dB/m, optical fibers have a bandwidth of tens of THz and loss as low as 0.17 dB/km [18].

Due to the fast growth of traffic, extreme amounts of optical fibers are deployed every year. The length of fiber optic cable shipped in 2011 surpassed the mark of 200 million kilometers [19]. This is more than the distance from our planet to our sun. The cost for fiber deployment is extremely high due to the construction work involved. Thus, there is a huge interest to increase the amount of data that can be transmitted on the already installed infrastructure.

This chapter will introduce the current limits when transmitting data on optical fibers and concepts that promise increased data rates on existing optical fiber links.

1.1 Available Bandwidth in Optical Transmission Systems

While optical fibers offer an optical bandwidth of several tens of THz (Fig. 1.2a), the usable bandwidth of optical fibers is mainly limited by the gain bandwidth of available optical amplifiers. The most commonly used optical amplifiers (erbium-doped fiber amplifiers - EDFA) provide gain within a bandwidth of up to 11 THz (see Fig. 1.2b – [20, 21]). Even though semiconductor optical amplifiers provide larger bandwidths, they typically add more noise than EDFAs and introduce additional wavelength crosstalk. Therefore, at the current state of development, semiconductor optical amplifiers are not suited for long transmission distances and systems with high powers and large numbers of wavelengths in wavelength division multiplexing (WDM).

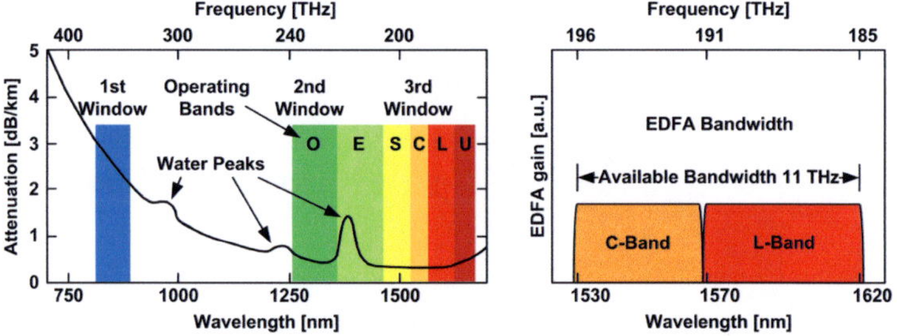

Fig. 1.2: Available bandwidth in optical networks. (a) Attenuation of optical fibers [22] and communication wavelength ranges with corresponding operating bands – total available bandwidth is larger than (b) the bandwidth of erbium doped fiber amplifiers, which spans the C and L-band.

This bandwidth limitation in state-of-the-art optical transmission leaves only two choices to increase data rates in an existing optical transmission system:

- Increasing the amount of data within the available optical bandwidth.

- Developing new amplifiers with a larger bandwidth.

Of these choices, the most attractive one is to increase the data rates within the available bandwidths as it requires no changes to the transmission link itself. The focus of this work is therefore on techniques to increase the information capacity in the available bandwidth.

A measure for the information capacity is the spectral efficiency (SE). The SE is defined as the transmitted bitrate BR (measured in bit/s) per required transmission bandwidth B:

$$SE = \frac{BR}{B} \tag{1.1}$$

Typically, optical transmitters have a bandwidth that is significantly smaller than the available fibre bandwidth. Therefore, wavelength division multiplexing (WDM) is employed to fill the available bandwidth in Fig. 1.2. In radio communication systems, this technique is known as frequency division multiplexing (FDM). Thus, there are two factors that can be optimized to increase the SE: First, the information capacity of the optical transmitters has to be increased. The information capacity is mainly determined by the modulation format. Second, the packing density of the channels from different transmitters has to be optimized. The packing density is determined by the multiplexing scheme chosen for the transmission system. The following sections will cover the state-of-the-art and advancements in modulation formats and multiplexing schemes for optical communication systems.

1.2 Modulation and Multiplexing

Modulation and multiplexing are two important principles of communication systems. Modulation is the process of encoding information for transmission over the medium. Multiplexing is the process of distributing resources of a communication channel to transmit several data streams at the same time. There are a number of physical dimensions that can be used for modulation and multiplexing [23], namely time, frequency, quadrature, polarization, space, and code. All these dimensions can be used for multiplexing and modulation. To achieve highest SE, the right combination of multiplexing and modulation schemes has to be selected. It is also important to consider that by modulating in a physical dimension, the signal occupies a certain space in this dimension that may no longer be used for multiplexing. Otherwise, crosstalk from one signal to the other will occur.

1.2.1 Optical Modulation Formats

In optical communication systems, advanced modulation formats have the potential to increase the spectral efficiency significantly, as will be discussed in the following sections. When modulating an optical carrier, one has the option to modulate four parameters, amplitude, phase, frequency, and polarization. The polarization is used for multiplexing in the present experiments. It will therefore not be included in the discussion of the modulation formats.

The most commonly used modulation format in optical communication systems has been on-off-keying (OOK). In recent years, differential phase shift keying (DPSK) has emerged as a format with large tolerances towards signal degradation by transmission effects [24]. Also, a first step towards an increase in spectral efficiency was made by the introduction of differential quadrature phase shift keying (DQPSK) [25] and DQPSK with polarization multiplexing [26].

These modulation formats were chosen due to the fact that the optical bandwidth was readily available and that they could be generated and received using simple hard-wired logic circuits. Until recently, the high data rates on optical channels made advanced signal processing nearly impossible due to the required processing speeds. However, processing speed is crucial to handle more advanced modulation formats like quadrature amplitude modulation (QAM).

In recent years, advanced modulation formats like quadrature amplitude modulation (QAM) have emerged [27]. In these modulation formats, the amplitude and phase of an optical carrier are modulated simultaneously, which allows to transmit several bit per transmission step or symbol. In case of 16QAM, 4 bit can be transmitted per symbol when using only one polarization. Recently, we have used 512QAM modulation to transmit 54 Gbit/s in an optical bandwidth of only 3 GHz. These advanced modulation formats have been well known in the community of wireless communication, but were not feasible for optical communication systems due to the amount of digital signal processing required. The exponential growth in digital signal processing speed has now finally reached a point, where real-time processing of such signals has become feasible [1, 3-6, 28-32].

1.2.2 Optical Multiplexing Schemes

This introduction in optical multiplexing schemes is based on the work presented in [8]. Here, it has been extended by a discussion of Nyquist WDM and polarization multiplexing to cover the schemes investigated in this work. The work on Nyquist WDM and Nyquist pulse shaping is based in [13] and [6], respectively. Because the following material partially stems from a Nature Photonics publication [8], both units, Tbit/s and Tbit s^{-1} are used synonymously.

1.2.2.1 State-of-the-Art in Optical Multiplexing Schemes

A combination of up to three multiplexing techniques is used in state-of-the-art transmission systems: Time division multiplexing (TDM), wavelength division multiplexing (WDM), and polarization division multiplexing (PolMUX).

Time division multiplexing is the traditional multiplexing scheme. Multiple bit streams are multiplexed onto a single signal by assigning a recurring timeslot to each of the tributaries. Due to the limited bandwidths of electronics, optical TDM (OTDM) was used to increase the data rates on a single laser further. Using OTDM, data rates of up to 10.2 Tbit s^{-1} were demonstrated using a single laser [33]. To achieve these data rates, TDM pulses are short and spectrally broad, see Fig. 1.3b, which makes OTDM schemes challenging. These short pulses (e.g. 300 fs [33]) require an extremely large receiver bandwidth that can only be achieved using optical sampling techniques. Also, the corresponding large optical bandwidth (e.g. 30 nm [33]) makes precisely engineered dispersion compensation a necessity.

Alternatively, with wavelength division multiplexing large aggregate data rates of up to 102.3 Tbit/s are transmitted over optical fibres [34, 35]. In WDM, the data are distributed onto several wavelengths to be transmitted in parallel. Therefore, optical pulse durations are longer and optical bandwidths per WDM channel are moderate (Fig. 1.3a). Spectral guard bands are required to avoid crosstalk from one channel to its neighbors. This reduces the spectral efficiency, and dedicated receivers and transmitters with stabilized lasers for each transmission wavelength are required.

In addition to these two techniques, polarization division multiplexing can be used to transmit two data streams on orthogonal polarizations.

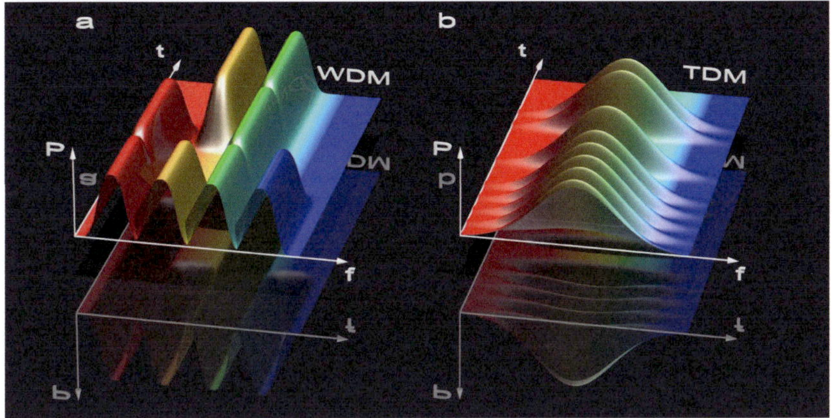

Fig. 1.3: Temporal and spectral evolution of traditional multiplexing schemes. (a) For wavelength division multiplexing (WDM), data are distributed over several carrier wavelengths (here four) and transmitted in parallel. The different wavelength tributaries can be separated by optical band-pass filters. (b) In time division multiplexing (TDM), data are transmitted in form of a serial stream of pulses. The plots show an illustration of the power (P) of the different formats over frequency (f) and time (t).

1.2.2.2 Advanced Multiplexing Schemes

Advanced multiplexing schemes can increase the information density within the available spectrum. This increase can be achieved by packing the information more densely in time or in frequency. When going to the theoretical limit, there are two distinct schemes, orthogonal frequency division multiplexing (OFDM) and Nyquist wavelength division multiplexing (Nyquist WDM). Both schemes are so-called super-channel approaches. Super-channels typically consist of several frequency-locked optical carriers onto which the data are encoded.

OFDM is a more recent approach in optical communications [7, 36-39], while already well known from wireless data transmission. In contrast to WDM, the modulated OFDM sub-carriers overlap with each other significantly, Fig. 1.4a. This way the spectral efficiency is high, and the encoding of information on a large number of subcarriers makes OFDM more tolerant to dispersion [38-40]. A detailed summary and discussion of advantages and disadvantages can be found [37]. In OFDM, the transmitted pulses are of a rectangular shape within an observation time interval. This leads to sinc-shaped spectra as illustrated in Fig. 1.4a, which are spaced such that the zero-field of all subcarriers coincides with the maximum field of an arbitrary subcarrier. Modulated OFDM subcarriers are orthogonal as shown in Section 2.2.1.1and Appendix A.1.1 [37].

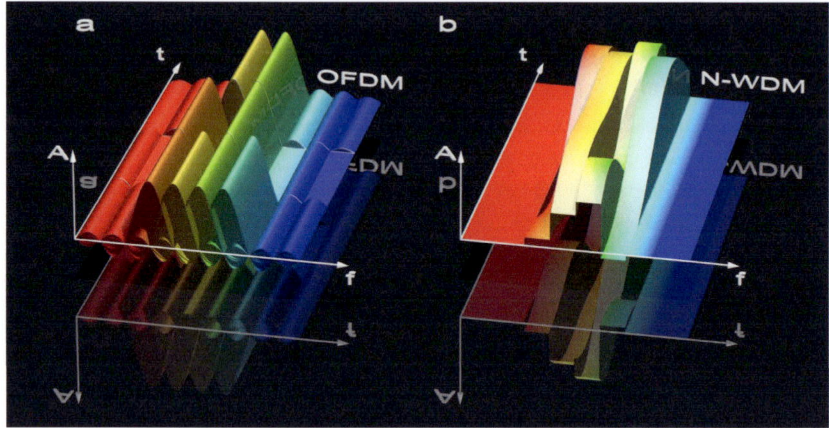

Fig. 1.4: Temporal and spectral evolution of advanced multiplexing schemes. (a) In an orthogonal frequency division multiplexing (OFDM) scheme, data are transmitted on a number of subcarriers in parallel. However, as modulated OFDM subcarrier spectra overlap, simple bandpass filters cannot separate the tributaries. Therefore, signal processing at the receiver with, e. g., a Fourier transform processing is required. (b) In a Nyquist WDM (N-WDM) scheme, modulated sinc-pulses are transmitted on the optical carriers. The spectrum of these sinc pulses is rectangular, which confines the spectrum of the modulated carriers to the Nyquist bandwidth of the signal. It is now possible to place the spectra next to each other without introducing crosstalk. The plots show an illustration of the evolution of the amplitude (A) of the different formats over frequency (f) and time (t).

Sinc-shaped Nyquist pulses are the key to the implementation of Nyquist WDM. These sinc-shaped pulses have a rectangular spectrum that is confined to the Nyquist bandwidth of the signal (Fig. 1.4b [5, 6]). For Nyquist WDM, the modulated signals can be placed next to each other without any guard band. As the Nyquist bandwidth is typically equal to the symbol rate of the signal, the carrier spacing of Nyquist WDM is also equal to the symbol rate. Nyquist WDM with higher order modulation formats like 16QAM has been discussed as an option for Tbit/s super-channels [41] as it offers a high spectral efficiency. Up to the experiments presented here, only a 400 Gbit/s demonstration has been shown [42]. In several other experiments, guard bands are used [43, 44].

However, for both schemes presented here, a key factor is the frequency stability of the subcarriers or carriers with respect to each other. The carrier generation is therefore quite challenging and is often implemented using optical frequency comb generation [7, 40, 45-49].

1.3 Summary

Optical communication systems are omnipresent and essential in science, economy, and our modern society. A tremendous effort is necessary to cope with the exponential growth in worldwide data traffic. Advanced modulation formats and multiplexing schemes show great promise for next generation communication systems, and are therefore of great research interest for future communication systems.

To increase the spectral efficiency in next generation communication systems, the following components and concepts have been investigated:

- A flexible optical multi-format transmitter, Chapter 3.

- Optical comb generation, Chapter 4.

- All-optical signal processing for OFDM, Chapter 0.

- Single-laser Tbit/s transmission, Chapter 6.

Chapter 2 contains an introduction into the theoretical background while additional material is provided in the Appendix.

2 Theoretical Background

This chapter covers the theoretical background on optical modulation formats and optical multiplexing schemes.

2.1 Modulation Formats

To transmit digital data over an optical link, the data has to be encoded onto an optical carrier. When neglecting the polarization, the electrical field of such a carrier $E(t)$ over time t can be described by its amplitude A, the carrier frequency f_c, and its phase ϑ.

$$E(t) = \underbrace{A}_{\text{Amplitude}} \cos(2\pi \underbrace{\times f_c \times t}_{\text{Frequency}} + \underbrace{\vartheta}_{\text{Phase}})$$

(2.1)

These three parameters can be modulated, which leads to the three fundamental modulation schemes: amplitude modulation, frequency modulation, and phase modulation. In digital systems with discrete states, these formats are usually referred to as amplitude shift keying (ASK), frequency shift keying (FSK), and phase shift keying (PSK). In the examples in Fig. 2.1a–c, only one bit is transmitted per symbol. Until recently, on-off-keying (OOK) was the most commonly used modulation format in optical communication systems, see Fig. 2.1d.

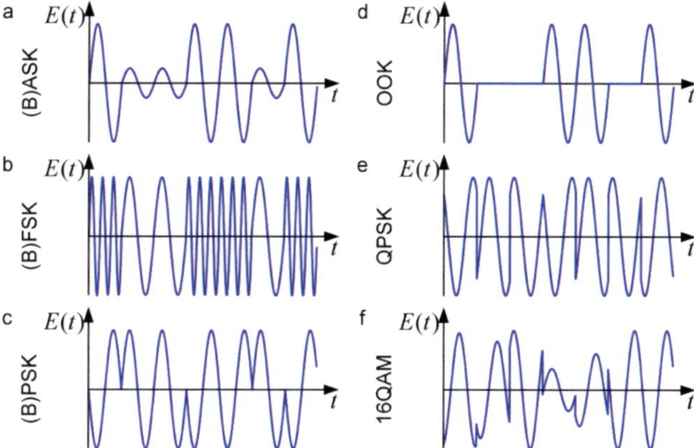

Fig. 2.1: Time functions of a carrier modulated with different modulation formats. (a) (Binary) amplitude shift keying, (B)ASK, (b) (binary) frequency shift keying, (B)FSK, (c) (binary) phase shift keying, (B)PSK, (d) on-off-keying, OOK, (e) quadrature phase shift keying, QPSK, and (f) 16-ary quadrature amplitude modulation, 16QAM.

Advanced modulation formats offer an increased amount of information per transmitted symbol. This increase is obtained by introducing additional possible states in the modulation format. The currently most popular advanced modulation format is quadrature phase shift keying (QPSK). In QPSK, data is encoded in four possible phase states (Fig. 2.1d).

By modulating more than one parameter of the carrier, the number of possible states can be increased even further. However, phase and frequency are closely related (e.g. a linear phase change over time is an effective frequency shift of the signal). So a combination of amplitude and phase modulation has become the most common scheme. Usually, quadrature amplitude modulation (QAM) is used when modulating phase and amplitude of an optical carrier. In quadrature amplitude modulation, equation (2.1) is transformed using the addition theorem:

$$
\begin{aligned}
E(t) &= A\cos\left(2\pi\, f_c\, t + \vartheta\right) \\
&= A\left(\cos\left(2\pi\, f_c\, t\right)\cos\left(\vartheta\right) - \sin\left(2\pi\, f_c\, t\right)\sin\left(\vartheta\right)\right) \\
&= \underbrace{A\cos\left(\vartheta\right)}_{I=\text{inphase}}\cos\left(2\pi\, f_c\, t\right) - \underbrace{A\sin\left(\vartheta\right)}_{Q=\text{quadrature}}\sin\left(2\pi\, f_c\, t\right) \\
&= I\cos\left(2\pi\, f_0\, t\right) - Q\sin\left(2\pi\, f_0\, t\right).
\end{aligned}
\tag{2.2}
$$

Here, the signal is represented by a superposition of an in-phase (I) component encoded on the original carrier $\cos\left(2\pi\, f_c\, t\right)$, and a quadrature ($Q$) component encoded on a carrier with a $\pi/2$ phase shift $\cos\left(2\pi\, f_0\, t + \pi/2\right) = -\sin\left(2\pi\, f_0\, t\right)$. The two carriers $\cos\left(2\pi\, f_c\, t\right)$ and $-\sin\left(2\pi\, f_0\, t\right)$ are orthogonal as seen in the orthogonality relation

$$
\int_0^{\frac{1}{f}} \cos\left(2\pi f\, t\right)\cdot\sin\left(2\pi f\, t\right)dt = 0.
\tag{2.3}
$$

Amplitude and phase of the signal $E(t)$ are given by

$$
A = \sqrt{I^2 + Q^2} \quad \text{and} \quad \vartheta =
\begin{cases}
\arctan\left(\dfrac{Q}{I}\right) & I \geq 0 \\[2ex]
\arctan\left(\dfrac{Q}{I}\right) + \pi & I < 0
\end{cases}.
\tag{2.4}
$$

The variables I and Q define a phasor

$$
\mathbf{A} = I + \mathrm{j}Q = A e^{\mathrm{j}\vartheta}.
\tag{2.5}
$$

The phasor $\mathbf{A}$ can be plotted in the constellation diagram (Fig. 2.2 – left hand side), where the real part is plotted on the I-axis and the imaginary part is plotted on the Q axis. A direct implementation of equation (2.2) yields the setup shown in Fig. 2.2 (center part). The phasor $\mathbf{A}$ is sometimes referred to as field vector or complex field vector, even though it is only a complex representation of amplitude and phase of the modulated carrier. When displaying a signal, the common choice is to display the phasors of the optimum sampling points of the signal and overlay them in one constellation diagram. With the constellation diagram, signals with advanced modulation formats like 16QAM (see time waveform in Fig. 2.1(f)) and 64QAM can be assessed much more easily, see Fig. 2.3.

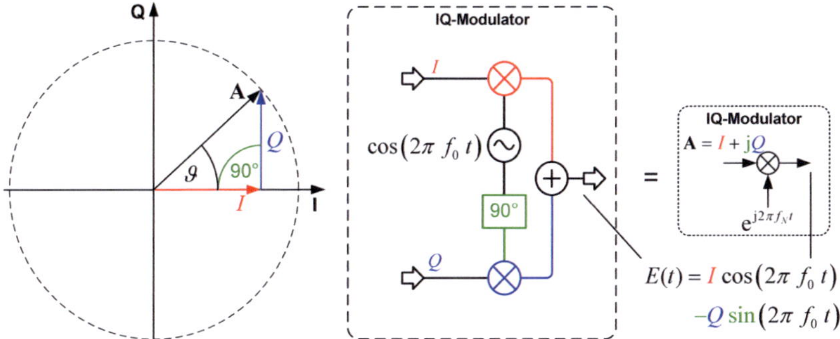

Fig. 2.2: Constellation diagram and IQ-modulation. The constellation diagram (left hand side) shows the complex representation **A** that is defined by in-phase I and quadrature Q component. Amplitude $A = |\mathbf{A}|$ and phase $\vartheta = \measuredangle \mathbf{A}$ can also be seen in the constellation diagram. Equation (2.2) directly gives the structure of the IQ-modulation scheme. The carrier $\cos(2\pi\, f_0\, t)$ is split, one part is modulated with the in-phase I component, the second part is shifted by $90°$ and modulated with the quadrature component Q. The IQ modulator is often plotted as a single complex modulator (right hand side).

The modulation of in-phase and quadrature allows transmitting complex data on a single optical carrier. The real part of the signal is encoded as in-phase component and the imaginary part is encoded as quadrature component. Due to the orthogonality of inphase and quadrature component as derived in eq. (2.3), the two components can be separated without crosstalk at the receiver.

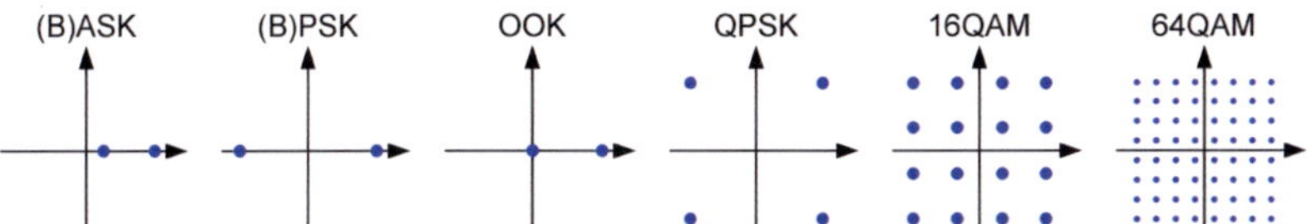

Fig. 2.3: Exemplary constellation diagrams are given for (B)ASK, (B)PSK, OOK, QPSK, 16QAM, and 64QAM. In these constellation diagrams, all possible field vectors ($\bullet$) of the signal at the sampling points are plotted in one diagram.

2.1.1 Modulation

To transmit data over a transmission link, it is modulated onto a carrier. This section covers several modulation schemes for signals with real-valued and complex modulation. The actual implementation of optical modulators is discussed in Appendix A.2.

2.1.1.1 Real-Valued Modulation

Real-valued modulation requires only a single modulator as shown in Fig. 2.4. Such a modulator can only change the in-phase component of the carrier. Therefore, only the amplitude can be modulated freely, while the resulting phase will be 0° for positive values of I and 180° for negative values of I. Modulation is therefore limited to constellation points on the I-axis in the constellation diagram. Typical modulation formats are on-off-keying (OOK) and binary phase shift keying (BPSK). BPSK is sometimes also referred to bipolar amplitude modulation.

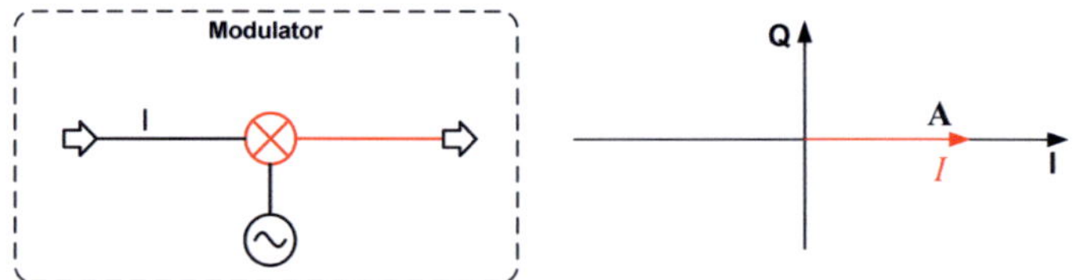

Fig. 2.4: A single modulator generates a signal modulated only in one quadrature. Only the I-axis of the constellation diagram can be accessed. The phase is limited to values of $0°$ and $180°$.

Fig. 2.5 shows the generation of OOK and BPSK in the complex domain. It illustrates the relation between in-phase signal (I – ——), quadrature phase signal (Q – ——), time signal (——) and the constellation diagram ($\bullet$). In case of real-valued modulation, the quadrature phase signal is always zero.

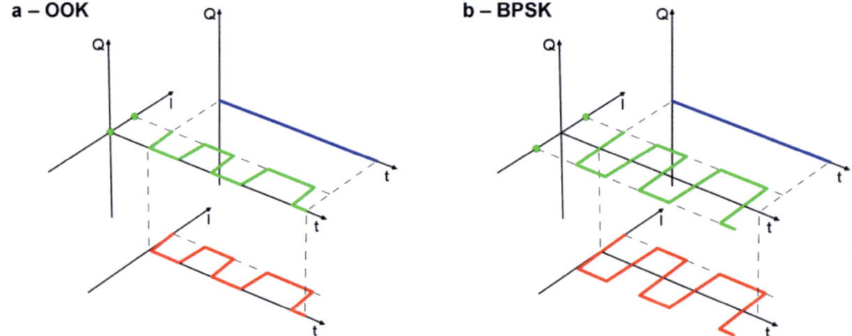

Fig. 2.5: Generation of a signal modulated in one quadrature. The modulator of Fig. 2.4 generates the in-phase (I, ——) signal. A quadrature phase (Q, ——) signal does not exist. The output signal (——) has only an in-phase component. The projection of the sampling points on the constellation plane on the left hand side generates the constellation diagram ($\bullet$). The constellation points are all placed on the I-axis. The signals are plotted over the time t. The concept is displayed for (a) on-off-keying (OOK) and (b) binary phase shift keying (BPSK).

2.1.1.2 Quadrature Amplitude Modulation

Quadrature amplitude modulation (QAM) is one of the most common complex modulation schemes. It is typically generated using the IQ-Modulator of Fig. 2.2 [50]. In this modulator, a signal with an arbitrary amplitude and phase (represented by the phasor $\mathbf{A}$, see eq. (2.5)) is synthesized through superposition of two orthogonal components (eq. (2.2) and (2.3)). These components have a $90°$ phase difference and can address the whole complex modulation plane, see Fig. 2.6. This scheme can generate any kind of modulation format depending only on the electrical drive signals.

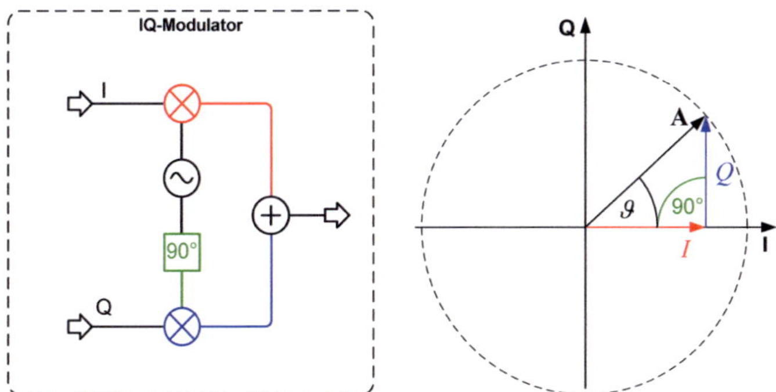

Fig. 2.6: The IQ-modulator generates an arbitrary field vector with the amplitude A and the phase ϑ by superposition of two copies of the carrier which are multiplied with amplitudes for in-phase (I) and quadrature phase (Q) and of which the carrier for the quadrature component is retarded by 90° in phase. The constellation diagram illustrates this process.

Similar to Fig. 2.5, the diagram Fig. 2.7 illustrates the generation of QPSK and 16QAM by an IQ-modulator. For QPSK, four phase states with the same amplitude are addressed with two two-level signals. For 16QAM, two four-level signals generate 16 states with 3 amplitudes and 12 phases.

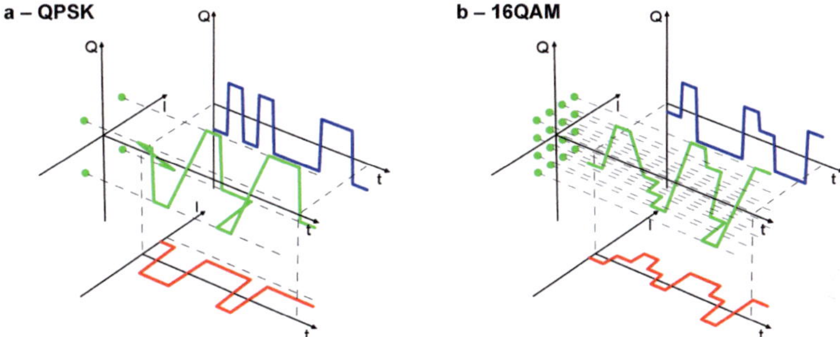

Fig. 2.7: Generation of modulation formats in an IQ-modulator. The two modulators in the IQ-modulator (Fig. 2.6) generate the in-phase (I, ——) and quadrature phase (Q, ——) signals. The superposition of these signals with a 90° phase shift generates the output signal (——). The projection of the sampling points on the constellation plane on the left hand side generates the constellation diagram (●). The concept is displayed for (a) quadrature phase shift keying (QPSK) and (b) 16ary quadrature amplitude modulation (16QAM).

2.1.2 Demodulation

After transmission, the format encoded on the optical carrier is demodulated in a receiver. This section covers the implementation of different receiver schemes that were used during the experiments presented in this thesis. The first two receiver schemes, direct and differential detection, are still the most used schemes in optical communication systems. The third technique, the coherent detection, is widely spread in the mobile communication world and is now being established in optical communication systems, due to the progress in the speed of modern semiconductor chips for digital signal processing.

2.1.2.1 Direct detection

Direct detection is the most basic receiver principle. It is commonly used at data rates of up to 40 Gbit/s. However, direct detection is insensitive to phase and frequency modulation of the received carrier and therefore only usable for intensity or amplitude modulated signals. Direct detection is the standard receiver concept for OOK modulation.

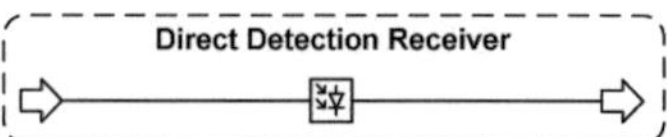

Fig. 2.8: Direct detection receiver concept. The intensity of the incident light is detected in a photodiode. This receiver is therefore insensitive to phase and frequency modulation.

The photodiode in the direct detection scheme generates an electrical current, which is proportional to the incident optical power P_e. As the power is proportional to the square of the electrical field, this detector is called square law detector. The generated current i is given by [51]

$$i = SP_e \; , \; S = \frac{\eta e}{hf_L} , \tag{2.6}$$

with the photo detector sensitivity S, also known as responsivity, the quantum efficiency η, the photon energy hf_L, and the elementary charge e.

2.1.2.2 Differential detection

Differential detection has been first established as a receiver concept for commercial systems with the demonstration of differential phase shift keying (DPSK) [24]. Among other things, differential phase shift keying offers an increased tolerance to noise, when compared to OOK [52, 53].

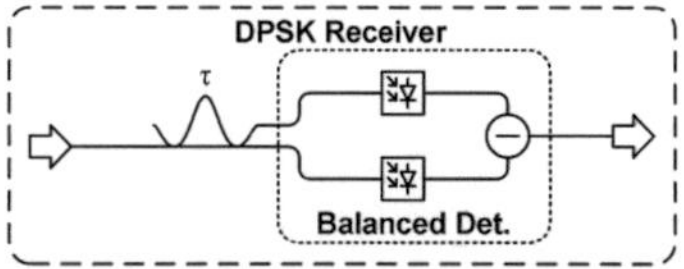

Fig. 2.9: Differential detection receiver for the detection of DPSK signals. The incoming signal is delayed by the time $\tau = 1/R$ defined by the reciprocal symbol rate to compare the phase of two consecutive symbols.

In the case of DPSK, the receiver comprises one delay interferometer [54, 55] and a balanced detector, see Fig. 2.9. The delay in the interferometer is usually equal to the symbol period of the signal $\tau = 1/R$. The delay interferometer effectively compares the phase of the signal at the time t to the phase of the signal that entered the interferometer at the time $t - \tau$.

However, this receiver cannot discriminate between a positive or negative phase shift (e.g. ϑ and $-\vartheta$). It is therefore not suitable for phase modulation with more than two phase states.

To discriminate between positive and negative phase changes, a second interferometer is required, Fig. 2.10. This interferometer includes an additional phase shift of 90° to discriminate +90 and −90° phase shifts.

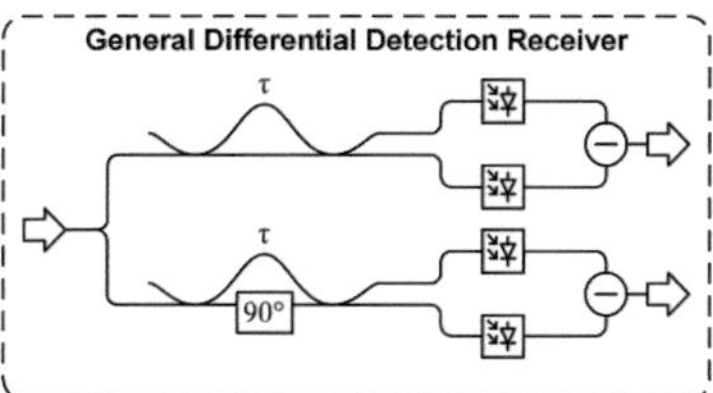

Fig. 2.10: General differential detection receiver. Introducing a second delay interferometer allows the receiver to discriminate between positive and negative phase changes.

In recent years, we extended the scheme by introducing a polarization diversity technique together with digital signal processing, so we could implement differential detection receivers for several modulation formats [56-58].

2.1.2.3 Coherent detection

To coherently receive a signal, it is mixed with a local oscillator (LO), see Fig. 2.11. In case of optical communication systems, the local oscillator is a laser. This mixing process generates frequency components at the sum frequency and at the difference frequency of LO and signal. In electronic receivers, the frequency components at the sum frequency are removed by a low pass filter. In optical communication systems, these filters are not necessary, as photo detectors only detect the difference frequency. Similar to the transmitter, the signal is mixed with the local oscillator and with a copy of the local oscillator that has been shifted by 90° in phase.

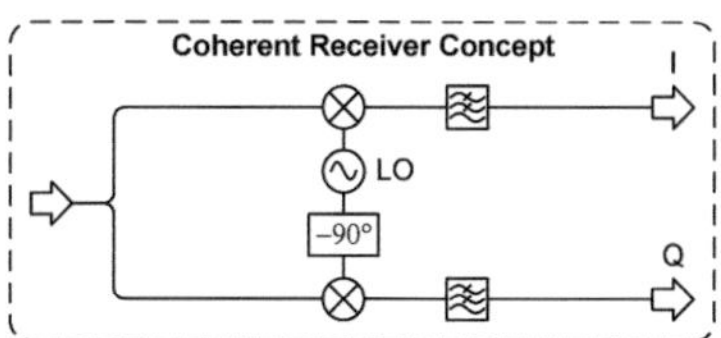

Fig. 2.11: Concept of a coherent receiver. In reverse to the process of the IQ modulator, the incoming signal is mixed with a local oscillator and a copy that is shifted by 90° in phase. A low pass filter is only required for the reception of electronic signals, as photo diodes in optical coherent receivers do not emit the sum frequency.

If phase and frequency of the local oscillator are locked to the carrier of the incoming signal, this scheme will directly recover the original signal. This is known as homodyne detection. However, in typical communication systems, the local oscillator is not locked to the carrier of the incoming signal. When the local oscillator frequency is within the incoming signal band, the receiver is a so-called intradyne receiver. In an intradyne receiver, digital signal processing is used to determine phase and frequency offset and to recover the original signal. If the local oscillator frequency is outside of the signal bandwidth, the lower arm of the coherent receiver can be omitted. This scheme is then called heterodyne reception. The implementation of a coherent receiver for optical signals is presented in Appendix A.3.

2.2 Multiplexing Schemes

Traditionally, two basic multiplexing schemes have been extremely popular in optical communication systems: wavelength division multiplexing (WDM) and time division multiplexing (TDM), see Fig. 1.3.

In WDM, the data are transmitted in parallel data streams. Each data stream is assigned a wavelength channel. These wavelength channels do not overlap in frequency and are separated by guard bands. Thus, they can be treated as completely independent channels. However, when using the typical guard bands, WDM is not that spectrally efficient.

In time division multiplexing, the data are multiplexed in a way that one high speed serial data stream is generated. Such a scheme is limited by the maximum speed of the transmitter. Optical time division multiplexing can be used to overcome speed limitations of electronics. Due to the pulse shapes in optical time division multiplexing, the bandwidth of such a channel is significantly larger than necessary.

Recently, polarization division multiplexing was introduced to increase the spectral efficiency in optical communication systems [59] and it is now being deployed in commercial systems. By transmitting two data streams on two orthogonal polarizations, the available bandwidth can be used twice, effectively doubling the data capacity available. Fig. 2.12 shows the resulting electrical field in the two polarizations of a dual polarization BPSK (DP-BPSK) signal and a dual polarization QPSK (DP-QPSK) signal.

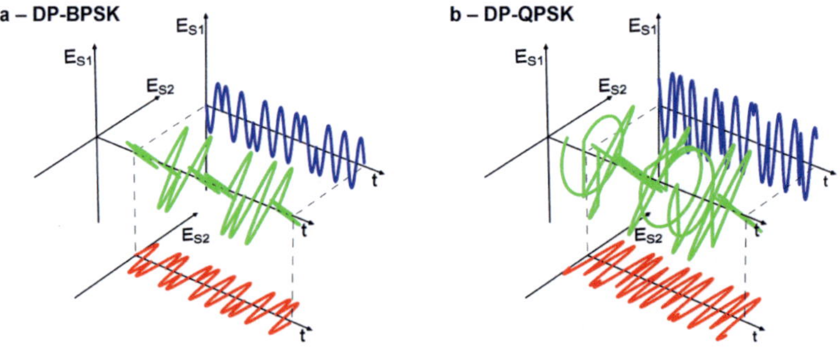

Fig. 2.12: Electrical field of polarization-multiplexed signals. The electrical fields in the two orthogonal polarizations S1 (———) and S2 (———) are superimposed and generate the output field (———). This is shown for dual polarization BPSK (DP-BPSK) and dual polarization QPSK (DP-QPSK) signals.

Advanced multiplexing schemes offer the possibility to enhance the spectral efficiency. This is achieved by making full use of the available spectrum. This can be achieved using orthogonal frequency division multiplexing (OFDM) or Nyquist WDM.

In case of OFDM, this is accomplished by allowing a spectral overlap of the modulated subcarriers. Due to this overlap, standard filters are no longer sufficient to separate the subcarriers at the receiver. A special receiver has to be designed that makes use of the orthogonality of the carriers.

2.2.1 Orthogonal Frequency Division Multiplexing (OFDM)

2.2.1.1 Principles of OFDM

The following section is based on a section in [60] and section 2.1 in [61]. A good reference for OFDM in optical communication systems is [37]. OFDM is a special case of a multi-carrier modulation (MCM) system where all carriers are orthogonal with respect to each other. In case of OFDM, the term subcarrier is used as a name for an OFDM carrier. The concept of a general MCM system is shown in Fig. 2.13.

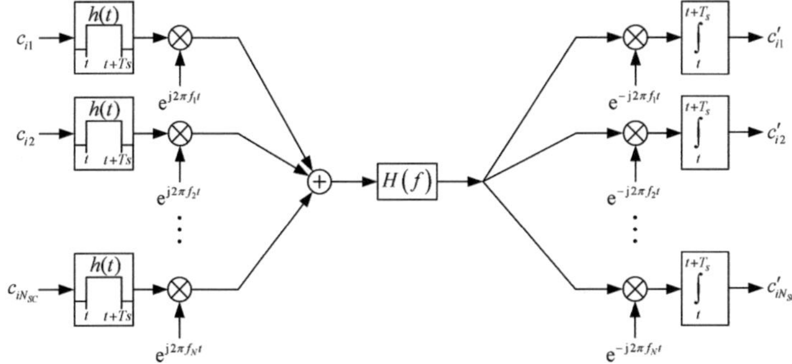

Fig. 2.13: Generic scheme for a multi-carrier modulation (MCM) system with N carriers. The data (c_{ik}, with $k = 1, 2 ..., N$) are encoded on rectangular pulses and modulated on multiple carriers $e^{j2\pi f_k t}$. The carriers are summed up to form the MCM signal. After transmission through the channel $H(f)$, the subcarriers are downconverted and the data c'_{ik} are recovered through an interate and dump filter.

To generate the OFDM signal, the baseband data c_{ik} are modulated onto the corresponding subcarriers $e^{j2\pi f_k t}$, with $k = 1, 2 ..., N$. The sum of all subcarriers is transmitted through a transmission link with the spectral impulse response $H(f)$ and split up at the receiver using matched correlators. The summed signal may be expressed as shown below.

$$x(t) = \sum_{i=-\infty}^{+\infty} x^{(i)}(t), \quad x^{(i)}(t) = \sum_{k=0}^{N-1} c_{ik} x_k(t - iT_s),$$

$$x_k(t) = \text{rect}\left(\frac{t}{T_s}\right) e^{j2\pi f_k t}, \quad \left|f_{k+1} - f_k\right| = F_s = \frac{1}{T_s}.$$

$$(2.7)$$

Here c_{ik} corresponds to the data in the ith symbol on the kth subcarrier with x_k being the subcarrier waveform. The subcarrier waveform is present for a given symbol length T_s and the data have to be constant for this duration. This is a fundamental requirement that should be kept in mind for the discussion of all-optical OFDM. The received signal which passed through an ideal channel is equal to the signal at the transmitter. After renaming variables for convenience, the signal at the receiver follows directly.

$$r\left(t+iT_s\right) = \sum_{i=-\infty}^{+\infty} \sum_{k=0}^{N-1} c_{ik} x_k\left(t\right) \tag{2.8}$$

To receive the signal on subcarrier k, a matched correlator is used. The output can be written as the correlation of the received signal (2.8) and the complex conjugate of the carrier waveform (2.7) of one symbol i taken over the symbol period T_s.

$$c'_{ki} = \frac{1}{T_s} \int_0^{T_s} r_i\left(t+iT_s\right) x_k^* dt \tag{2.9}$$

Using equation (2.8) it can be seen that the output of the kth channel is the auto-correlation integral times the symbol c_{ik} plus the sum of the cross-correlations with all other subcarriers.

$$c'_{ik} = \frac{1}{T_s} \int_0^{T_s} c_{ik} x_k x_k^* dt + \frac{1}{T_s} \int_0^{T_s} \sum_{l \neq k} c_{il} x_l x_k^* dt \tag{2.10}$$

The cross-correlation terms will degrade the signal quality and should therefore be zero after the correlator. Looking at two carriers k and l yields the condition of vanishing cross-correlation terms for a particular receiver channel.

$$\frac{1}{T_s} \int_0^{T_s} x_l x_k^* dt = \frac{1}{T_s} \int_0^{T_s} e^{j\left(2\pi\left(f_l-f_k\right)t\right)} dt = e^{j\left(\pi\left(f_l-f_k\right)T_s\right)} \operatorname{sinc}\left(\pi\left(f_l-f_k\right)T_s\right) \overset{!}{=} \delta_{lk} \tag{2.11}$$

This condition is also known as the orthogonality relation for an OFDM signal. The solution of the integral is the product of an exponential function and a sinc-function. Since the exponential function will never be zero, the sinc-function must be zero in order to make (2.11) vanish. This condition is satisfied for the differences in frequency between carrier l and carrier k being an integer multiple of the reciprocal symbol length.

$$f_l - f_k = m\frac{1}{T_s}, \tag{2.12}$$
$$m \in \mathbb{Z}$$

The plot in Fig. 2.14 shows four orthogonal subcarriers modulated with a symbol length of T_s. The sinc-functions in frequency originate from the rectangular windowing of the modulated subcarriers with a width of T_s.

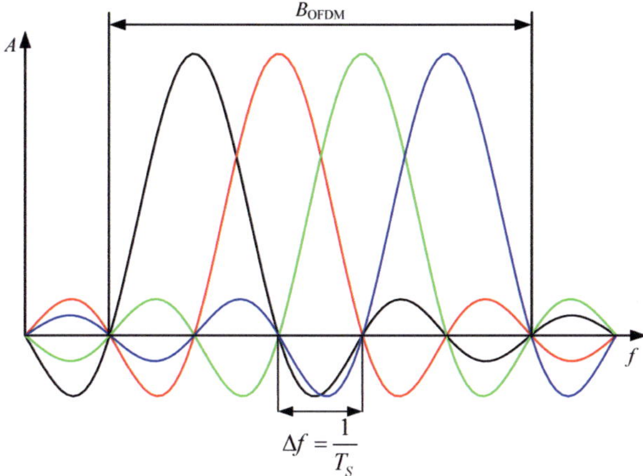

Fig. 2.14: Spectrum of four modulated OFDM subcarriers x_k. The zero points of the sinc-spectra coincide with the position of the neighboring carriers.

Implementing a MCM system as described above using individual modulators and correlators for each channel is very costly. It is only practical for extremely large signal bandwidths where digital signal processing is not powerful enough. For signals with a smaller bandwidth, OFDM signals are generated using digital signal processing.

2.2.1.2 Generation of OFDM with Digital Signal Processing

Low speed OFDM signals are usually generated using digital signal processing [37]. Today's electronics is capable of processing signals for data rates of up to 100 Gbit/s [4] in real-time. Fig. 2.15 illustrates the most basic concept for the digital processing of an OFDM signal in the transmitter and the receiver.

To generate an OFDM signal through digital signal processing, the data to be encoded on the OFDM subcarriers are interpreted as complex Fourier coefficients of the signal. The inverse discrete Fourier transform of these Fourier coefficients yields complex time samples of a signal that contains the data as modulation on its subcarriers. Usually, the inverse discrete Fourier transform is implemented as an inverse fast Fourier transform (IFFT). A parallel-to-serial conversion and a digital-to-analog conversion (DAC) generate the actual OFDM signal.

To receive an OFDM signal, the signal is converted into the digital domain in an analog-to-digital converter (ADC). A serial-to-parallel conversion supplies the complex time samples to the discrete Fourier transform or fast Fourier transform (FFT). The FFT recovers the complex Fourier coefficients of the signal that are the data transmitted.

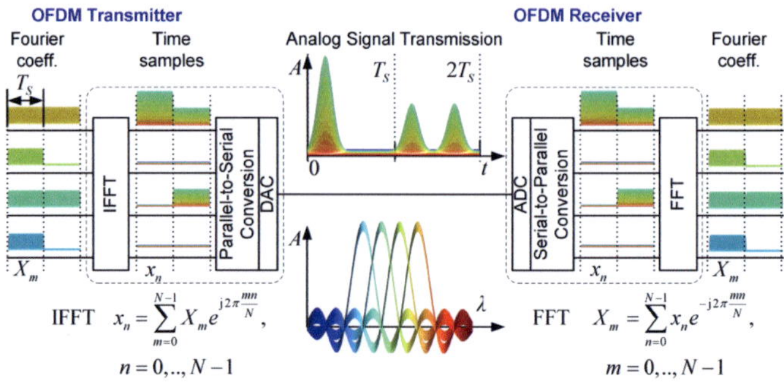

$$\text{IFFT}\quad x_n = \sum_{m=0}^{N-1} X_m e^{j2\pi \frac{mn}{N}},$$

$$n = 0,..,N-1$$

$$\text{FFT}\quad X_m = \sum_{n=0}^{N-1} x_n e^{-j2\pi \frac{mn}{N}},$$

$$m = 0,..,N-1$$

Fig. 2.15: Classical implementation of OFDM processing from Fig. 2 in [8]: Here, N = 4 parallel data-streams are interpreted as complex Fourier coefficients X_m in the frequency domain. The IFFT yields time-samples x_n of a signal. Parallel-to-serial and digital-to-analogue conversion provides an analogue OFDM signal A(t) with spectrum A(λ). After transmission, the information is retrieved through the inverse process: An analogue-to-digital conversion (ADC) and a serial-to-parallel conversion followed by a FFT return the Fourier coefficients representing the data streams.

It should be mentioned that a complex signal to be transmitted has to be modulated onto a carrier. This step has been omitted in this illustration. Also, in a real OFDM receiver, several additional synchronization steps are required. These steps are discussed in detail in [37]. A more detailed discussion of OFDM can be found in Chapter 0.

2.2.1.3 Cyclic prefix or Guard Interval

An OFDM signal as defined in Section 2.2.1.1 is only defined within the duration T_s. No definition has been made outside of this window. Also, OFDM symbol do not necessarily have to be transmitted directly adjacent to each other. This offers the opportunity to introduce additional signal components or guard intervals in between OFDM symbols as needed in a given transmission system. Here, two different situations are discussed, in which this is beneficial: In case of dispersion and for an implementation with discrete modulators that have a limited bandwidth. In both cases, the reason is that the complex data c_{ik}, which is transmitted on an OFDM subcarrier has to be constant within the same interval T_s at the receiver. Otherwise, the orthogonality relation (eq. (2.11)) is violated and the crosstalk (eq. (2.10)) is no longer negligible.

In case of dispersion, the different subcarriers experience a different group delay as illustrated in Fig. 2.16. To be able to tolerate this, a cyclic prefix or cyclic postfix, abbreviated with CP, is added to each subcarrier, which has the length of the maximum group delay difference. Different subcarriers still experience their respective group delay, however, there will always remain a window of T_s without phase and amplitude changes on all subcarriers. The different group delay translates into a phase shifts of the modulated data, which can be measured and compensated for.

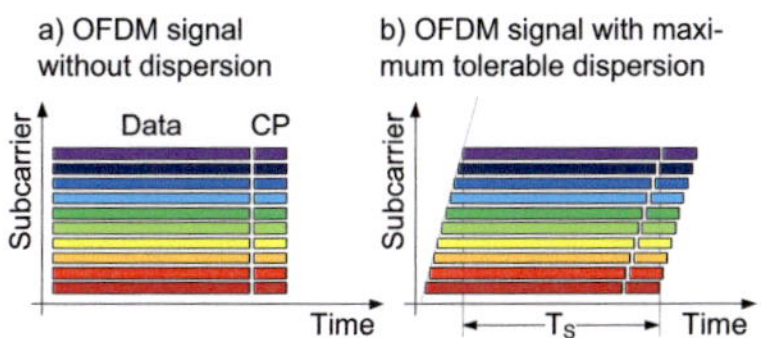

Fig. 2.16: Mitigating dispersion by means of a cyclic prefix (CP). (a) An OFDM signal with data and cyclic prefix. (b) Frequency dependent delay of subcarriers after transmission due to dispersion. The amount of dispersion that can be tolerated is limited by the length of the cyclic prefix (CP) as all subcarrier symbols must stay constant within T_s. This figure and caption have been published in [10].

In case of bandwidth limited modulators, rise and fall-times of these modulators can have a significant impact on the orthogonality. Therefore, a guard interval τ_{GI} or guard time is introduced in between symbols, which ensures that transitions take place outside of the integration interval T_S. This way, the orthogonality of the modulated subcarriers is maintained. This is discussed in more detail in Section 5.1.3.2.

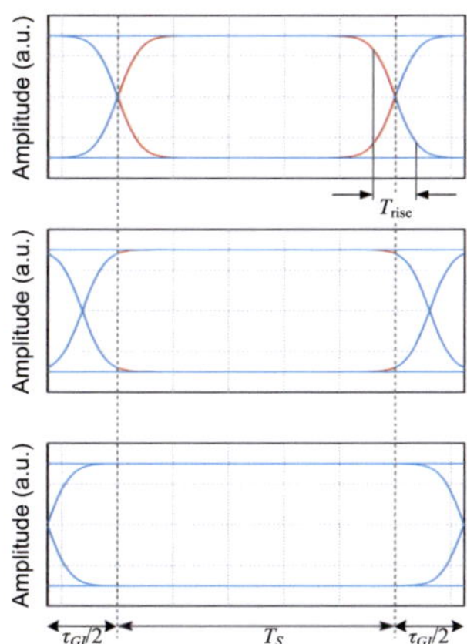

Fig. 2.17: Function of an OFDM guard interval. The exemplary eye diagram of a single modulated subcarrier is shown. In this illustration, the amplitude in arbitrary units corresponds to the modulated data c_{ik} of the OFDM signal, which must be constant during an OFDM symbol. If no guard interval is used, the symbol duration equals the integration interval T_S and the signal transitions, described by the 10-90% rise/fall time T_{rise} (red), violate the orthogonality relation (eq. (2.11)), which causes inter- and intra-subchannel crosstalk. Therefore, a guard interval guard interval τ_{GI} is introduced. The effect of this is discussed in detail in Section 5.1.3.2. This figure and caption are based on work published in [10].

2.2.1.4 OFDM Schemes in Optical Communication Systems

Several different OFDM schemes are discussed for optical communication systems. To illustrate the different concepts, only the most fundamental parts of the transmitters are shown, see Fig. 2.18 and Fig. 2.19. Two different types of OFDM transmitters are covered here: four schemes with electronic generation of OFDM and two schemes for the all-optical generation of OFDM.

22   Theoretical Background

The four electronics based OFDM transmitters are shown in Fig. 2.18. The most basic
scheme is an analog implementation of an OFDM transmitter, where the OFDM subcarriers
are each encoded separately (Fig. 2.18a). However, DSP-based OFDM schemes as discussed
in 2.2.1.2 are much more common (Fig. 2.18b-d). In general, the IFFT output data is complex
valued. It is therefore not possible to transmit it directly. One option and the simplest scheme
to generate a real-valued OFDM signal is through Hermitian symmetric extension of the data
(Fig. 2.18b), which will ensure that the IFFT output is real-valued. This real-valued baseband
OFDM signal is then encoded on an optical carrier. While the use of electronic mixers or IQ-
modulators is avoided, the required amount of digital signal processing is increased significantly. The other option is to use IQ-modulation on an electrical or optical carrier to transmit
the complex output of the IFFT. When using an electrical carrier, the scheme is often referred
to as "electrical" OFDM (Fig. 2.18c). After encoding the signal on the electrical carrier, it is
being modulated on the optical carrier. In case of IQ-modulation on an optical carrier for so-
called "optical" OFDM (Fig. 2.18d), the in-phase and quadrature data are encoded directly on
the optical carrier.

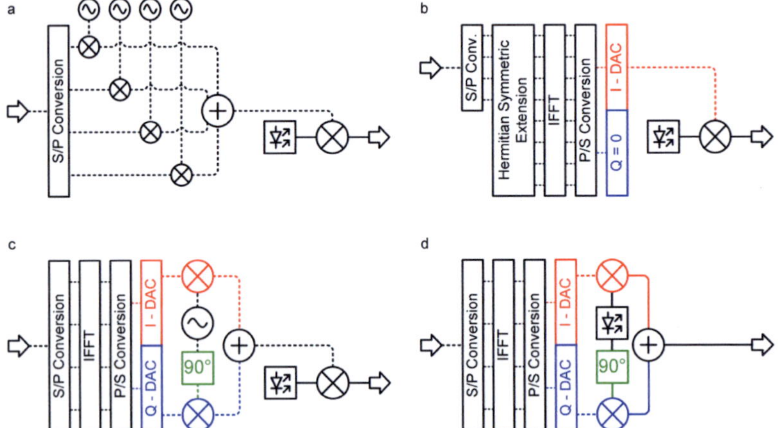

Fig. 2.18: OFDM transmitter schemes for optical communication systems. (a) Analog electrical
OFDM. After serial-to-parallel conversion (S/P conversion), the parallel data are encoded $\otimes$ on
electrical carriers. An addition $\oplus$ of the modulated carriers generates the OFDM signal to be encoded on an optical carrier. (b) DSP-based real-valued OFDM. After S/P conversion, a Herimitian
symmetric extension ensures that the signal after the IFFT has only real values. This way, a single
digital-to-analog converter can generate the analog signal. The signal is then modulated on an optical carrier (c) DSP-based electrical OFDM. The OFDM signal is generated through S/P conversion, IFFT, parallel-to-serial conversion (PS), and digital-to-analog conversion for in-phase and
quadrature (I-DAC and Q-DAC). It is modulated on an electrical carrier to generate a real valued
signal and subsequently modulated onto an optical carrier. (d) DSP-based optical OFDM. The signal is generated as in (c), however, it is directly encoded on an optical carrier in an optical IQ-
modulator.

For the signals in Fig. 2.18a-c, simple intensity or amplitude modulation suffices for
transmission of the signal and the quality requirements for the laser source are not extremely
high. Even directly modulated lasers can be used, which makes this system extremely inter-

esting for low cost applications. In case of intensity or amplitude modulation, a direct detection receiver is sufficient for reception of the signal. However, one could also choose to modulate the phase of the optical carrier with such a signal. The scheme in Fig. 2.18d increases the spectral efficiency significantly at the price of increased quality requirements for the optical carrier and an increased receiver complexity, as a phase and amplitude sensitive receiver is now mandatory. It is therefore most suited for applications, where the cost is not that significant and the increased performance and spectral efficiency are desired.

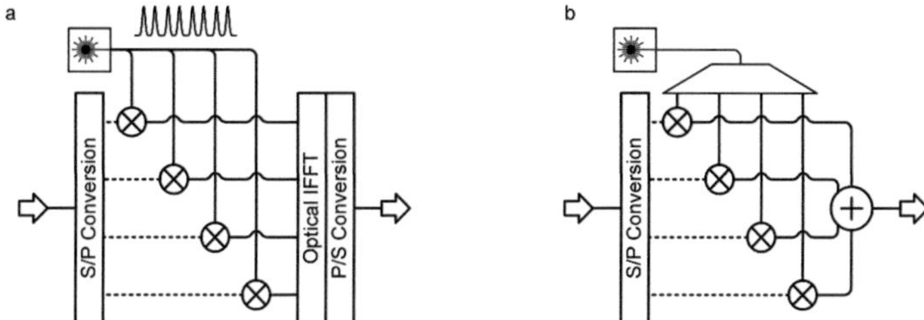

Fig. 2.19: All-optical OFDM transmitter schemes. (a) IFFT-based implementation of all-optical OFDM transmitter. The data are parallelized and modulated onto a pulsed laser source. An optical IFFT and parallel-to-serial conversion generate the OFDM signal. (b) Analog implementation of all-optical OFDM. The data are parallelized and modulated on the desired number of subcarriers. Here, these subcarriers are generated in an optical comb source. Subsequently, the modulated subcarriers are combined to form the OFDM signal. The symbol $\otimes$ represents IQ-modulators. Electrical connections are shown as dashed lines, while optical connections are solid lines.

Two schemes for the optical generation of OFDM are presented (Fig. 2.19a, b). The first scheme uses the optical IFFT, while the second scheme is an analog implementation of an OFDM transmitter. In the IFFT-based scheme (Fig. 2.19a), the data parallelized and encoded on optical pulses from an optical pulse source. Subsequently, the Optical IFFT and parallel-to-serial conversion generate the OFDM signal. The difficulty here lies in the insertion of a cyclic prefix. To this end, one would have to copy a certain part of the generated OFDM signal and insert it at another time. The difficulty here lies in the optical phase relation between the copy and the original part that has to be maintained. This scheme is discussed in more detail in Section 5.1.3.1. The second scheme is the analog implementation of the OFDM transmitter (Fig. 2.19b). Here, the data are parallelized and encoded on the respective subcarrier frequencies. These subcarrier frequencies are usually derived from a common source in a comb generator, which inherently generates equidistant carriers. After modulation, the modulated carriers are combined to form the OFDM signal. This scheme offers full flexibility when inserting a cyclic prefix by simply reducing the symbol rate of the data, effectively introducing a guard interval. The same effect can be achieved by an increase of the carrier spacing. It should be mentioned that in contrast to the other scheme, this scheme required a guard interval to accommodate the rise- and fall-times of the modulators. Additional information on this scheme can be found in Section 5.1.3.2.

2.2.2 Nyquist–Wavelength Division Multiplexing

The following section on Nyquist pulse shaping is an excerpt of [6].

Real–time Nyquist pulse generation beyond 100 Gbit/s and its relation to OFDM

R. Schmogrow,   M. Winter,   M. Meyer,   A. Ludwig,   D. Hillerkuss,
B. Nebendahl, S. Ben-Ezra, J. Meyer, M. Dreschmann, M. Huebner, J. Becker,
C. Koos, W. Freude, and J. Leuthold

Optics Express 20(1), 317-337 (2012). [6]

Abstract: Nyquist sinc-pulse shaping provides spectral efficiencies close to the theoretical limit. In this paper we discuss the analogy to optical orthogonal frequency division multiplexing and compare both techniques with respect to spectral efficiency and peak to average power ratio. We then show that using appropriate algorithms, Nyquist pulse shaped modulation formats can be encoded on a single wavelength at speeds beyond 100 Gbit/s in real-time. Finally we discuss the proper reception of Nyquist pulses.

2.2.2.1 Introduction

Sinc-shaped Nyquist pulses spread into adjacent time slots, but their rectangularly shaped spectra require only the minimum Nyquist channel bandwidth. They are well known from communication theory but are relatively new in optical communications. The Nyquist modulation format is very similar to optical orthogonal frequency division multiplexing (OFDM), where sinc-shaped sub-spectra extend into adjacent frequency slots, and symbols in time are rectangularly shaped. In the course of this paper all Nyquist pulses are sinc-shaped.

Here, we first discuss the close relation of Nyquist pulse modulation with OFDM [8, 61, 62]. Both Nyquist pulse shaping and OFDM are described with a similar formalism. This way it will become clear that Nyquist modulation is nothing but an orthogonal time division multiplexing technique, much the same as OFDM is an orthogonal frequency multiplexing technique. Furthermore, we compare the two multiplexing methods with respect to their characteristics like spectral efficiency (SE) and peak-to-average power ratio (PAPR). We then demonstrate real-time Nyquist pulse generation for signals beyond 100 Gbit/s. This has become possible even with the limited speed of state-of-the art electronics [32]. In more detail, we generate quadrature phase shift keying (QPSK) at 56 Gbit/s and quadrature amplitude modulation with 16 states (16QAM) at 112 Gbit/s in combination with polarization division multiplexing (PDM). This results in an overall spectral efficiency of 7.5 bit/s/Hz for PDM-16QAM. Finally the reception of Nyquist shaped pulses is discussed comparing it with the reception of standard non-return-to-zero (NRZ) QAM signals.

2.2.2.2 Advanced filtering in optical WDM networks

Modern optical networks rely on multi-wavelength and multi-carrier transmission systems in order to fully exploit the bandwidth offered by optical fibers. The ultimate target is to maximize the spectral efficiency, i. e., the amount of transmitted data within a given bandwidth [63]. In general, the maximum capacity of a channel is only limited by Shannon's law. For optical communications non-linear distortions limit the ultimate channel capacity at high launch powers. Thus increasing the signal to noise ratio (SNR) by increasing the signal power is only possible within certain limits [64, 65]. For high capacity networks, coming close to this so-called non-linear Shannon limit is of special interest.

For conventional M-ary QAM signals, the spectral occupancy does not alter significantly when changing the number of bits b transmitted per symbol. Thus increasing the number of constellation points $M = 2^b$ leads directly to an increase in spectral efficiency. However, transmitting an additional bit per symbol implies doubling the number of constellation points, so that for a constant average power the required SNR increases significantly. This is also true if the spectral efficiency is increased by polarization division multiplexing (PDM) or polarization switching [66].

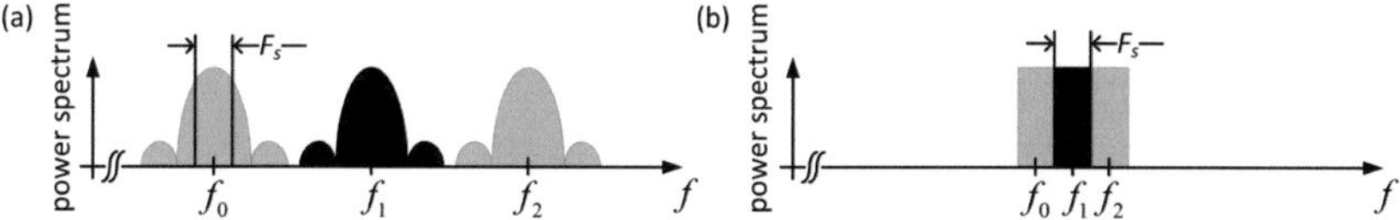

Fig. 2.20: Reducing the required channel spacing in WDM systems by removing spectral portions outside the Nyquist bandwidth F_s through filtering. Minimum channel spacing for crosstalk free systems is fixed by F_s. (a) Unfiltered M-ary QAM channels with theoretically infinitely wide spectrum. (b) Nyquist channel center frequencies can be spaced apart by the Nyquist bandwidth F_s. Basically, no guard bands are required.

Bandwidth can be saved, however, when applying advanced filtering. From signal theory we know that the minimum bandwidth needed to fully encode a bandwidth-limited signal is the Nyquist bandwidth F_s [67]. If the optimization of spectral efficiency is the ultimate target, all frequency components outside the Nyquist band must be removed by filters. As a consequence, the time domain signal changes from pulses that are clearly separated in time (e. g., non-return-to-zero format, NRZ) to pulses that overlap their neighbors.

As an example, Fig. 2.20(a) displays the spectrum of an M-ary QAM NRZ signal for three different WDM channels centered at optical frequencies f_0, f_1, and f_2. The spectra are significantly wider than F_s, but can be reduced to the Nyquist bandwidth without losing any signal information. However, appropriate filtering is required to achieve the best possible transmission quality. The sinc-shaped spectrum of an NRZ signal should be filtered such that the resulting spectrum is of rectangular shape under the assumption that the frequency response of the channel is flat in the region of interest. Therefore the side lobes must be removed, and the spectrum within F_s must be flattened. If there are slopes in the channel's frequency response, or the noise accumulated in the system is not constant over frequency, a pre-

and de-emphasis filtering scheme should be applied. In a properly filtered WDM spectrum comprising the same three carrier wavelengths as in Fig. 2.20(a), the channels can now be placed next to each other located on a frequency grid the minimum spacing of which is dictated by the symbol rate F_s (Fig. 2.20(b), Nyquist WDM 1[13, 41]).

In general, the previously described filters can be implemented optically, electrically, or digitally. Possible implementations are shown for a software-defined transmitter, Fig. 2.21. Optical filters with a transfer function $S_{21}(f)$ as in Fig. 2.21(a) could be used. The difficulty is to build optical filters with frequency responses that drop significantly inside just a few MHz. Optical filters based on liquid crystals may offer an opportunity to perform such filtering [68]. Nevertheless, these filters are quite elaborate and show some penalties due to the limited slopes in their frequency response. Electrical filters as shown in Fig. 2.21(b) are another option. They can provide very steep slopes. A complex transmitter, however, requires two filters with a specific frequency transfer function depicted in Fig. 2.21(b). As before, these analog electrical filters are not easily available. Conversely, designing digital filters to be included in the digital signal processing (DSP) part of the transmitter [69] seems to be a suitable option to solve the problem. State-of-the-art software-defined optical transmitters [3] utilize DSP functionality which only has to be extended. Naturally, digital filtering calls for additional analog anti-aliasing filters to remove image spectra. These filters can be standard low-pass filters as any negative influence can be pre-compensated by digital filtering. Furthermore, only DSP offers the flexibility to vary filter coefficients during runtime and therefore the capability to adapt to changes in the channel response. Additionally, changing the symbol rate F_s of the digital filter based transmitter is achieved without changing the hardware. Analog filters are generally fixed with respect to their frequency responses and cannot easily be altered.

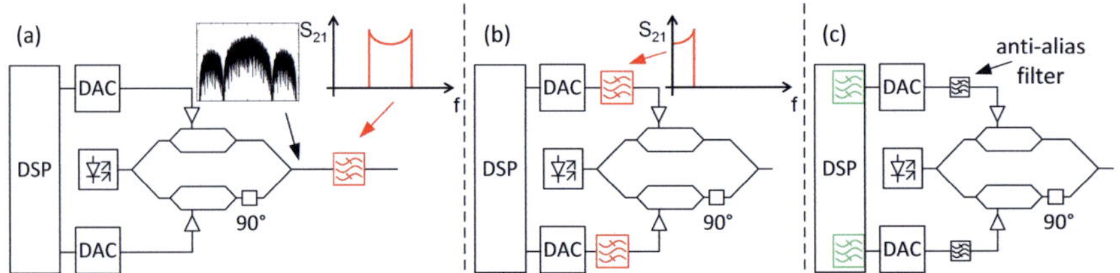

Fig. 2.21: Pre-filtering of optical M-QAM signals for generating sinc-shaped Nyquist pulses. (a) An optical filter can be applied to carve an ideal rectangle out of an NRZ spectrum. Such an optical filter requires a frequency response $S_{21}(f)$ that would be difficult – if not impossible – to realize. (b) Alternatively, two analog electrical filters can be used to form an appropriate output signal with an ideal rectangularly shaped spectrum. These filters would also need a non-standard transfer function $S_{21}(f)$. (c) By digital filtering as a part of the digital signal processing (DSP) block in the transmitter one can efficiently remove redundant parts of the spectrum. Resulting signals show almost ideal rectangular spectra, and additional off-the-shelf analog electrical or optical filters can easily remove spurious spectra.

2.2.3 Nyquist pulse modulation and OFDM: A comparison

The following section shows the relation of Nyquist pulse shaping and OFDM. This discussion is an extract of [6].

Real–time Nyquist pulse generation beyond 100 Gbit/s and its relation to OFDM

R. Schmogrow,   M. Winter,   M. Meyer,   A. Ludwig,   D. Hillerkuss,
B. Nebendahl, S. Ben-Ezra, J. Meyer, M. Dreschmann, M. Huebner, J. Becker,
C. Koos, W. Freude, and J. Leuthold

Optics Express 20(1), 317-337 (2012). [6]

Nyquist pulse modulation can be derived from the well-known optical orthogonal frequency division multiplexing (OFDM) technique [61]. This is done by simply interchanging time and frequency domain when describing the signal.

In general, an OFDM signal $x(t)$ is an *infinite* sequence of *temporal* symbols $x^{(i)}(t)$ superscripted with i. Each temporal symbol consists of a superposition of N *temporal sinusoidals* with equidistant carrier frequencies f_k inside a temporally rectangular window defining the temporal symbol length T_s. Frequency spacing $F_s = f_{k+1} - f_k = 1 / T_s$ and temporal symbol length T_s are interrelated to establish orthogonality, Eq. (8.2) in the Appendix. To simplify the discussion, we let aside a possible cyclic prefix that would reduce the symbol rate below F_s, and would therefore increase the temporal symbol spacing to a value larger than T_s. The OFDM carriers are encoded with complex coefficients c_{ik}. We find for the OFDM signal

$$x(t) = \sum_{i=-\infty}^{+\infty} x^{(i)}(t), \quad x^{(i)}(t) = \sum_{k=0}^{N-1} c_{ik} x_k (t - iT_s),$$

$$x_k(t) = \text{rect}\left(\frac{t}{T_s}\right) e^{j2\pi f_k t}, \quad |f_{k+1} - f_k| = F_s = \frac{1}{T_s}. \tag{2.13}$$

The rectangular function rect(z) is 1 for $|z| < 1/2$ and zero otherwise, see Eq. (8.1) in the Appendix. By Fourier transforming Eq. (2.13) we obtain the frequency domain representation of the i-th temporal OFDM symbol, i. e., a set of N spectral sinc-functions centered at frequencies f_k,

$$X(f) = \sum_{i=-\infty}^{+\infty} X^{(i)}(f), \quad X^{(i)}(f) = \sum_{k=0}^{N-1} c_{ik} X_k(f) e^{-j2\pi f iT_s},$$

$$X_k(f) = T_s \, \text{sinc}\left(\frac{f - f_k}{F_s}\right). \tag{2.14}$$

In contrast to OFDM, the *spectrum* $Y(f)$ of a Nyquist signal is a *finite* sequence of N *spectral* symbols superscripted with i. Each spectral symbol consists of a superposition of *infinitely many spectral sinusoidals* with equidistant Nyquist pulse position times t_k ("carrier" positions) inside a spectrally rectangular window defining the spectral symbol length F_s. Temporal spacing $T_s = t_{k+1} - t_k = 1 / F_s$ and spectral symbol length F_s are interrelated to establish orthogonality, Eq. (8.3) in the Appendix. The Nyquist "carriers" are again encoded with complex coefficients c_{ik}. In analogy to Eq. (2.13) we find

$$Y(f) = \sum_{i=0}^{N-1} Y^{(i)}(f), \quad Y^{(i)}(f) = \sum_{k=-\infty}^{+\infty} c_{ik} Y_k(f - iF_s),$$

$$Y_k(f) = T_s \, \text{rect}\left(\frac{f}{F_s}\right) e^{-j2\pi f t_k}, \quad \left|t_{k+1} - t_k\right| = T_s = \frac{1}{F_s}. \tag{2.15}$$

By Fourier transforming Eq. (2.15) we obtain the time domain representation of the ith spectral Nyquist symbol, i. e., a set of infinitely many temporal sinc-functions centered at times t_k,

$$y(t) = \sum_{i=0}^{N-1} y^{(i)}(t), \quad y^{(i)}(t) = \sum_{k=-\infty}^{+\infty} c_{ik} y_k(t) e^{+j2\pi i F_s t},$$

$$y_k(t) = \text{sinc}\left(\frac{t - t_k}{T_s}\right). \tag{2.16}$$

The relations Eq. (2.13) − (2.16) are visualized in Fig. 2.22. The left column, Fig. 2.22(a) and (c), describes the time-frequency correspondence for OFDM, while the right column, Fig. 2.22(b) and (d), relates to Nyquist pulses. The upper rows of each section in Fig. 2.22 show the time dependency of the signals, while the lower rows refer to the corresponding spectra.

In Fig. 2.22(a), three temporally sinusoidal subcarriers modulated with $c_{i1} = c_{i2} = c_{i3} = 1$ form a specific OFDM symbol with width T_s and positioned at $t = 0$. The OFDM spectrum is a superposition of three spectral sinc-functions located at frequencies f_{k-1}, f_k, and f_{k+1}, which are separated by F_s. In Fig. 2.22(b), the superposition of three temporal sinc-functions is seen which are located at times t_{k-1}, t_k, and t_{k+1} and separated by T_s. These Nyquist pulses are modulated with $c_{i1} = c_{i2} = c_{i3} = 1$ and form a specific spectral Nyquist symbol with width F_s and position at $f = 0$. It consists of three spectrally sinusoidal Nyquist "subcarriers". The graphs in Fig. 2.22(a) and (b) represent Eq. (2.13)–(2.16) for $i = 0$, i. e., for an OFDM and a Nyquist symbol positioned at $t = 0$ and $f = 0$, respectively.

If we set $k = 0$, then each of the three OFDM or Nyquist symbols shown here consists of only one temporal zero-frequency ($f_0 = 0$) or spectral zero-time ($t_0 = 0$) "sinusoidal", respectively. For OFDM, the three temporal symbols are positioned at times $(i-1)T_s$, iT_s and $(i+1)T_s$, Fig. 2.22(c). The resulting spectrum is located within a sinc-shaped envelope having its first zeros at $-F_s$ and $+F_s$. Due to the different positions of the temporal symbols we see three spectral sinusoidals within the (green) spectral envelope. For Nyquist pulses, the three temporal sinusoidals inside the (green) sinc-shaped pulse envelope with zeros at $-T_s$ and $+T_s$ correspond to three spectral symbols positioned at frequencies $(i-1)F_s$, iF_s, $(i+1)F_s$, Fig. 2.22(d).

A schematic of OFDM signal and Nyquist pulse generation is given in Fig. 2.23. The left column, Fig. 2.23(a) and (c), refers to OFDM, whereas the right column, Fig. 2.23(b) and (d), describes Nyquist pulse generation. For a better understanding we set one of the summation variables k or i of Eq. (2.13)–(2.16) to zero while varying the other one, and we present the signal generation in both frequency and time domain.

For OFDM signal generation in the frequency domain, Fig. 2.23(a), a real sinc-shaped spectrum $X_{k=0}(f)$ centered at $f=0$ is shifted by a finite number of equidistant frequency steps kF_s, $k=0...N-1$. The resulting sub-spectra are modulated by complex coefficients c_{ik}. The total OFDM spectrum $X^{(0)}(f)$ for $i=0$ is formed by superimposing all N subcarrier spectra (Σ stands for summation), resulting in an OFDM symbol located at $t=0$ only.

For Nyquist pulse generation in the time domain, Fig. 2.23(b), a real sinc-shaped impulse $y_{k=0}(t)$ centered at $t=0$ is shifted by an infinite number of equidistant time steps kT_s, $k=-\infty...+\infty$. The impulses are modulated by complex coefficients c_{ik}. The total Nyquist pulse $y^{(0)}(t)$ for $i=0$ is formed by superimposing all "subcarrier" pulses, resulting in a Nyquist pulse sequence at one carrier "frequency" $f=0$ only.

For OFDM pulse generation in the time domain, Fig. 2.23(c), a real rect-shaped pulse $x_{k=0}(t)$ comprising only one carrier "frequency" $f=0$ is shifted by an infinite number of equidistant time steps iT_s, $i=-\infty...+\infty$. These sub-pulses are modulated by complex coefficients c_{ik}. The total OFDM time signal $x(t)$ for $f=0$ is formed by superimposing infinitely many temporal sub-pulses.

For Nyquist signal generation in the frequency domain, Fig. 2.23(d), a real rect-shaped spectrum $Y_{k=0}(f)$ comprising only one Nyquist pulse ("carrier") at $t=0$ is shifted by a finite number of equidistant frequency steps iF_s, $i=0...N-1$. The resulting sub-spectra are modulated by complex coefficients c_{ik}. The total Nyquist symbol $Y(f)$ at $t=0$ is formed by superimposing all N sub-spectra.

OFDM and Nyquist *receivers* can be built similar to the transmitter scheme depicted in Fig. 2.23. To this end, the received signal would enter from the right, the symbol Σ would represent a splitter, and local oscillators with complex conjugate time dependency (OFDM signal) or complex conjugate Nyquist pulses (Nyquist signal) mix with the incoming signals to recover the modulation coefficients c_{ik} having integrated over the symbol period T_s (for OFDM signal) or over all times (for Nyquist signals). Forming the complex conjugate means reverting the signs of frequency steps F_s and time steps T_s, respectively.

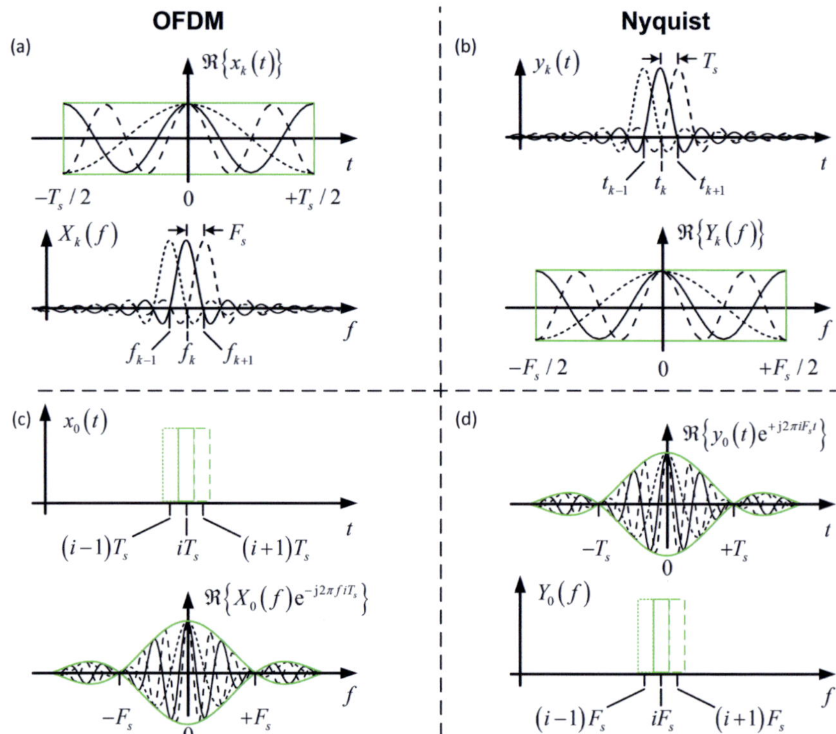

Fig. 2.22: Comparison between OFDM (left column) and Nyquist sinc-pulses (right column) in time and frequency domain. (a) The upper graph shows the real parts of three on-off keyed sinusoidal subcarriers $x_k(t)$, the sum of which represents one specific time-domain OFDM symbol centered at $t = 0$. The lower graph shows the corresponding spectra $X_k(f)$ of the subcarriers centered at frequencies f_{k-1}, f_k, and f_{k+1}. (b) The upper graph shows three Nyquist pulses $y_k(t)$ ("temporal subcarriers") centered at times t_{k-1}, t_k, and t_{k+1}. The lower graph shows the corresponding real parts of the spectra $Y_k(f)$ with three spectral sinusoidals, the sum of which represents one specific frequency-domain Nyquist symbol centered at $f = 0$. (c) The upper graph shows the envelopes (green rectangles as in Fig. 2.22(a) of three temporal OFDM symbols located at times $(i-1)T_s$, iT_s, and $(i+1)T_s$. For simplicity, each temporal OFDM symbol is composed of the same single zero-frequency subcarrier $f_0 = 0$ for $k = 0$, The lower graph shows the corresponding real parts of the spectra within a sinc-shaped envelope (green). The three spectral sinusoidals within correspond to three temporal positions of the temporal OFDM symbol. (d) The upper graph shows the real parts of three Nyquist pulses within a sinc-shaped envelope (green). The three temporal sinusoidals within correspond to three spectral positions of the spectral Nyquist symbols. The lower graph shows the envelopes (green rectangles as in Fig. 2.22(b) of the three spectral Nyquist symbols located at frequencies $(i-1)F_s$, iF_s, and $(i+1)F_s$. For simplicity, each spectral Nyquist symbol is composed of the same single zero-time Nyquist "subcarrier" $t_0 = 0$ for $k = 0$.

An in-depth mathematical comparison between OFDM and Nyquist pulse shaping is given in the Appendix. Due to the close relation to OFDM, Nyquist pulse generation could be also referred to as an orthogonal time division multiplexing technique.

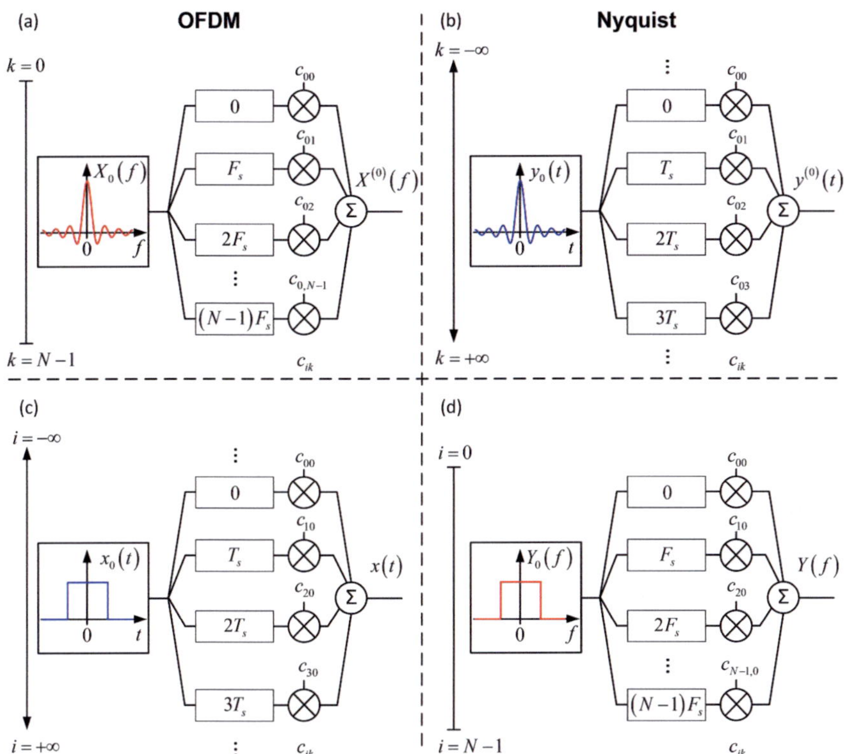

Fig. 2.23: Schematic of OFDM signal and Nyquist pulse generation, both in frequency domain and time domain. For a better understanding we keep either k or i of Eq. (2.13)–(2.16) constant at zero while varying the other quantity. (a) OFDM spectrum, OFDM symbol at $t_0 = 0$ ($i = 0$) only: N sinc-shaped sub-spectra for a finite number of $k = 0\ldots N-1$ keyed subcarriers are frequency shifted by kF_s and modulated with complex coefficients c_{ik}. The superposition of all modulated sub-spectra results in the total OFDM spectrum $X(f) = X^{(0)}(f)$. (b) Nyquist pulse, Nyquist symbol with $f_0 = 0$ ($i = 0$) only: Infinitely many sinc-shaped pulses (Nyquist "subcarriers", $k = -\infty\ldots+\infty$) time shifted from $t = 0$ by increments kT_s and modulated with complex coefficients c_{ik}. The superposition of all modulated pulses results in the total Nyquist time signal $y(t) = y^{(0)}(t)$. (c) OFDM symbol, OFDM spectrum with $f_0 = 0$ ($k = 0$) only: Infinitely many rect-shaped temporal pulses ($i = -\infty\ldots+\infty$) time shifted from $t = 0$ by increments iT_s and modulated with complex coefficients c_{ik}. The superposition of all modulated pulses results in the total OFDM time signal $x(t)$. (d) Nyquist symbol, Nyquist pulse at $t_0 = 0$ ($k = 0$) only: N rect-shaped sub-spectra for a finite number of $i = 0\ldots N-1$ keyed spectral symbols are frequency shifted by iF_s and modulated with coefficients c_{ik}. The superposition of all modulated spectra results in the total Nyquist spectrum $Y(f)$.

2.3 Digital Signal Processing

This section introduces some building blocks for digital signal processing (DSP) that are used throughout this thesis.

2.3.1 Pseudo Random Bit Sequences (PRBS)

Pseudo random bit sequences (PRBS) are digital test sequences specified by the International Telecommunication Union (ITU) in [70]. Such test sequences are required for bit error testing of communication systems as they emulate the random properties of real traffic as closely as possible. Well defined test sequences are required to perform true bit error ratio (BER) measurements.

Pseudo random bit sequences are the most commonly used test sequences. A PRBS has a length of $2^n - 1$ and contains a maximum number of n-1 consecutive zeros and n consecutive ones, with n being a natural number. A PRBS is generated using a linear feedback shift register. A linear feedback shift register for the generation of a PRBS $2^7 - 1$ is shown in Fig. 2.24.

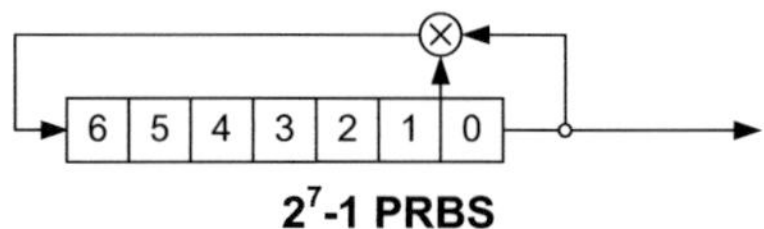

Fig. 2.24: Linear feedback shift register that generates a pseudo random bit sequence with a length of $2^7 - 1$ [60]. In this shift register, values for certain positions in the shift register are combined with an XOR ($\otimes$) and fed back to the input of the shift register.

A detailed description of the generation of a PRBS at highest bit rates can be found in Section A.6.6.

2.3.2 Fast Fourier Transform

This section on the fast Fourier transform has been published in [10] and is also included in Section 5.1.1. The figure has been adapted to comply with the definition of symbols in this work. The implementation and application of an optical FFT is discussed in detail in Chapter 5.1.

The fast Fourier transform (FFT) is an efficient method to calculate the discrete Fourier transform (DFT) for a number of time samples N, where $N = 2^p$ with p being an integer. The N-point DFT is given as

$$X_m = \sum_{n=0}^{N-1} \exp\left[-\mathrm{j}2\pi\frac{mn}{N}\right] x_n \ , \quad m = 0,..,N-1 \tag{2.17}$$

transforming the N inputs x_n into N outputs X_m. If the x_n represent a time-series of equidistant signal samples of signal $x(t)$ over a time period T, as shown in Fig. 2.25(a), then the X_m will be the unique complex spectral components of signal x repeated with period T [71]. The FFT typically "decimates" a DFT of size N into two interleaved DFTs of size $N/2$ in a number of recursive stages [72] so that

$$X_m = \begin{cases} E_m + \exp\left[-\mathrm{j}\dfrac{2\pi}{N}m\right]O_m & \text{if } m < \dfrac{N}{2} \\[3mm] E_{m-N/2} - \exp\left[-\mathrm{j}\dfrac{2\pi}{N}\left(m-\dfrac{N}{2}\right)\right]O_{m-N/2} & \text{if } m \geq \dfrac{N}{2} \end{cases} \tag{2.18}$$

The quantities E_m and O_m are the *even* and *odd* DFT of size $N/2$ for even and odd inputs x_{2l} and x_{2l+1} ($l = 0,1,2,\ldots N/2{-}1$), respectively. Fig. 2.25(b) shows the direct implementation of the FFT for $N = 4$ using time electrical sampling and signal processing.

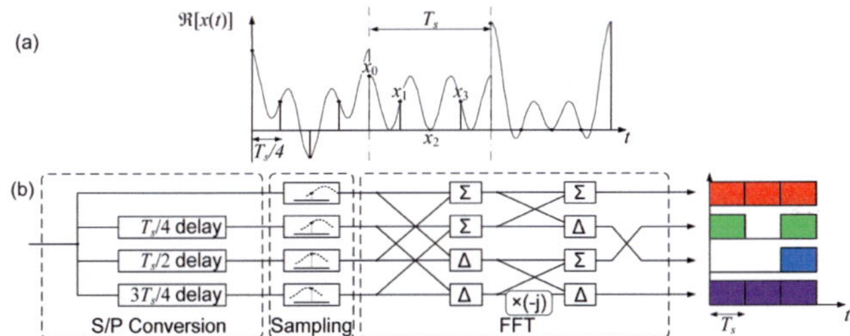

Fig. 2.25. Four point example of the traditional fast Fourier transform. (a) Exemplary signal in time sampled at $N{=}4$ points; (b) the structure consists of a serial-to-parallel (S/P) conversion that generates parallel samples of the signal $x(t)$ that are delayed by an integer multiple of $T_S/4$, a sampling stage to generate the time samples x_n and a conventional FFT stage that calculates the fast Fourier transform of the sampled signal. The right-hand side of (b) shows typical output signals for input signal (a).

2.3.3 Digital–to–Analog Converter Basics

Digital-to-analog converters (DAC) are needed to generate the analog waveforms of signals generated in digital signal processing. DACs are therefore an essential part of our multi-format transmitter.

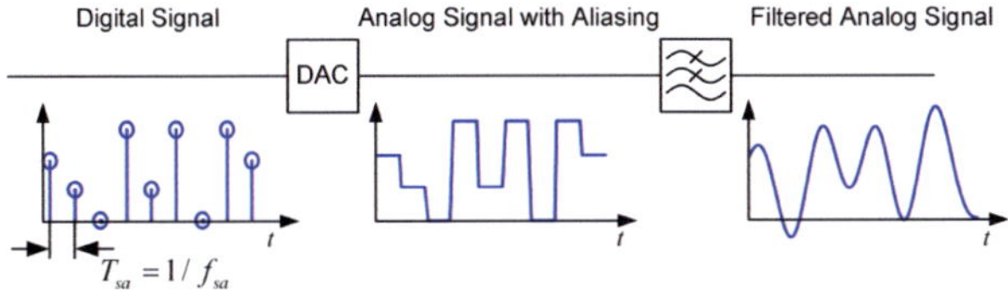

Fig. 2.26: Digital-to-analog converter and anti-aliasing filter. The digital signal with a sample spacing of $T_{sa} = 1/f_{sa}$ is converted to the analog domain in the DAC. A DAC typically has a rectangular impulse response which leads to the illustrated analog signal. To generate a signal without aliasing, a low pass filter removes all frequencies higher than f_{max}.

Key figures of a DAC are the resolution N_R in bit and the sampling rate f_{sa} in Samples/s (Sa/s). The resolution is the number of bit per sample, which defines the number of different levels 2^{N_R} a DAC can generate. Digital signals are only defined at discrete points in time that are separated by $T_{sa} = 1/f_{sa}$, (Fig. 2.26 left inset). The maximum frequency f_{max} that can be present in the digital input signal is given by the sampling theorem [73]:

$$f_{max} = f_{sa}/2 \tag{2.19}$$

The rectangular impulse response of the DAC leads to the signal illustrated in Fig. 2.26. If desired, alias frequencies can be removed in an anti-aliasing filter. However, it depends strongly on the application if such filters are desired. In case of the all-optical OFDM concept, these filters have to be omitted, as fast rise- and fall-times are desired (see Section 2.2.1.4). When employing pulse shaping for Nyquist WDM, such filters are needed to avoid any aliasing frequencies that would fall into neighboring frequency bands.

3 Flexible Optical Multi-Format Transmitter

Flexible optical multi-format transmitters are prerequisite for next generation adaptive communication links that can adapt to changing data rate demands and channel quality.

Additionally, in contrast to today's standards, where a different kind of transmitter is built for each kind of modulation format, such a flexible optical multi-format transmitter promises a significant reduction in manufacturing and maintenance costs as only one type of transmitter can be used to replace all transmitters.

This chapter was published in part on the conference on signal processing in photonic communications [1] and in [60]. This multi-format transmitter offers online signal generation with data rates up to 168 Gbit/s. The transmitter generates a number of modulation formats ranging from on-off-keying (OOK) to 16 quadrature amplitude modulation (16QAM) at up to 30 GBd. The choice of modulation formats is limited only by the physical limits of the hardware and the programming of the field programmable gate arrays (FPGA).

3.1 Introduction

Software-defined multi-format transmitter with real-time signal processing for up to 160 Gbit/s

D. Hillerkuss, R. Schmogrow, M. Huebner, M. Winter, B. Nebendahl, J. Becker, W. Freude, and J. Leuthold

in *Signal Processing in Photonic Communications* (Optical Society of America, 2010), paper SPTuC4. [1]

Advanced optical transmitters are required to implement advanced modulation formats like phase shift keying (PSK) [74] and quadrature amplitude modulation (QAM) [50] that are of growing interest for optical communication systems [75]. Additionally, advanced multiplexing schemes like orthogonal frequency division multiplexing (OFDM) [37, 76] promise increased data rates at improved spectral efficiencies for next generation optical transmission systems. However, the commonly used transmitter structures for the various advanced modulation formats and multiplexing schemes differ significantly [33, 63, 77, 78], so that each newly implemented modulation format needs different transmitters and receivers. In contrast, software-defined transmitters (and receivers) are much more flexible because the modulation format is defined by the programming of the transmitter. In addition to the multitude in modulation formats that can be generated with such a transmitter it will also allow for signal pre-distortion [77, 79] to compensate transmitter imperfections and mitigate transmission impairments. The price for this flexibility is the requirement for high-speed digital-to-analog converters (DAC).

DAC-based transmitters have been used previously for a successful implementation of higher-order modulation formats [63]. However, those setups either achieved only a low symbol rate like 4 GBd [63], suffered from a low DAC resolution and high complexity due to the use of discrete components for the implementation of the DAC [80], or require special optical modulators [77]. Recently, high-speed DACs have emerged that allow for symbol rates beyond 25 GBd. A predistorting transmitter at 10.7 Gbit/s has been already demonstrated [81].

In this chapter, we demonstrate the first software-defined transmitter, which is capable of generating on-off keying (OOK), binary and quadrature phase shift keying (BPSK, QPSK), and 16QAM signals at symbol rates up to 30 GBd.

3.2 Optical Multi–Format Transmitter Concepts

The concept chosen for the implementation of an optical transmitter has major impact on its versatility and performance. A summary of different schemes possible is displayed in Fig. 3.1.

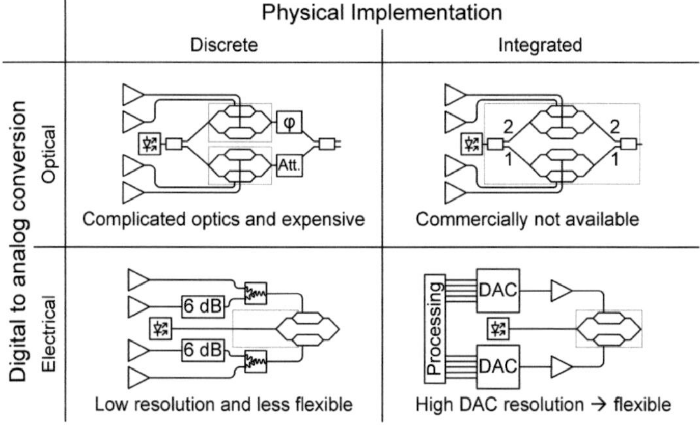

Fig. 3.1: Possible concepts for the implementation of an optical multi-format transmitter. It is necessary to choose between discrete and integrated implementations and between optical and electrical digital to analog conversion.

A decision to be made is if a discrete or an integrated solution is more feasible and if the digital to analog conversion should be implemented in the optical or in the electrical domain.

Typically one would prefer an optical digital to analog conversion, as the nonlinearity of the Mach-Zehnder interferometer can be exploited to reduce noise present on the electrical drive signals of the modulators. This way, the requirements for electrical signal to noise ratio (SNR) can be relaxed, or the signal quality can be improved for a constant SNR. However, a discrete optical implementation is not feasible due to the required phase stabilization in our example (Fig. 3.1 upper left), and integrated modulators are currently only exist as prototypes [77].

As an optical digital to analog conversion is not feasible, the only choice is between discrete and integrated digital to analog conversion. While highest sampling rates are currently achieved with discrete implementations [82], the resolution of discrete implementations is currently limited to a few bit [83]. The resolution is critical for advanced modulation formats as it defines the achievable SNR due to quantization noise. Due to these reasons, an integrated DAC was chosen for our implementation of the multi-format transmitter.

3.3 Transmitter Design

Section 3.3 on the transmitter design is based on work presented in [60]. Details on the Implementation of the Transmitter can be found in Appendix A.6.

> ### *Implementation of an OFDM Capable Optical Multi-Format Transmitter*
>
> R. Schmogrow, Master Thesis No. 803, *Institute of Photonics and Quantum Electronics* (Karlsruhe Institute of Technology, Karlsruhe, 2009). [60]

3.3.1 Setup

The transmitter setup shown in Fig. 3.2 consists of several components including digital-to-analog-converters (DAC), field programmable gate arrays (FPGA), an optical I/Q modulator as described in A.2.2, and a computer to control and supervise the involved devices. Fig. 3.2 illustrates the setup as well as data and control flow. The sampling clock is provided by a radio frequency synthesizer (RF-synthesizer) and split to supply the two DACs. The clock phase of the DACs is adjusted using an RF phase shifter.

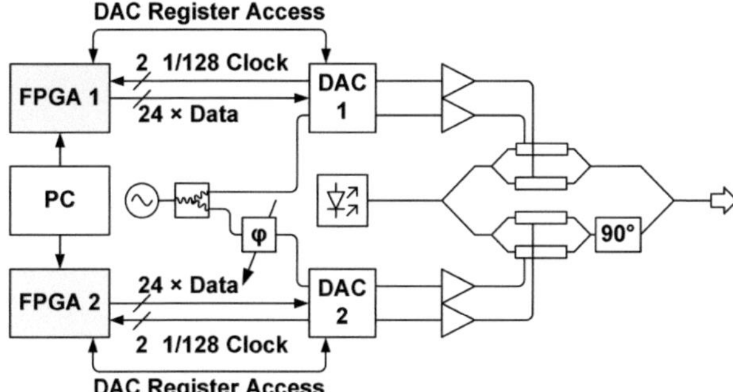

Fig. 3.2: Multi-format transmitter implementation. The signal is generated in real-time on two field programmable gate arrays (FPGA). The FPGAs supply the data to high speed digital to analog converters (DAC). The analog output of the DACs is amplified and drives an optical IQ-modulator in which the signal is modulated onto an optical carrier. A computer provides control functionalities for the different components of the transmitter.

The PC transmits waveforms and control information directed to the FPGAs. The FPGAs are used to drive two DACs that can operate at up to 34 GSa/s. For prototyping, waveforms of advanced complex modulation schemes such as OFDM can be created offline on the PC and loaded into the memory located on the FPGA boards. The stored sequences are run repeatedly and modulated onto an optical carrier in the transmitter. Once tested, these formats can be implemented in real-time on the FPGAs. On off keying (OOK), phase shift keying (PSK) and quadrature amplitude modulation (QAM) are demonstrated here.

3.3.2 Field Programmable Gate Array Architecture

The FPGA architecture is chosen carefully in order to create a highly flexible system that can be extended easily. Each of the components should be replaceable without changing the whole structure of the FPGA. A processor is used for all control and synchronization tasks that do not require highest speeds, whereas other parts that require high speed operations are implemented in hardware using the very high speed integrated circuit (VHSIC) hardware description language (VHDL). Fig. 3.3 shows the architecture implemented on the FPGAs.

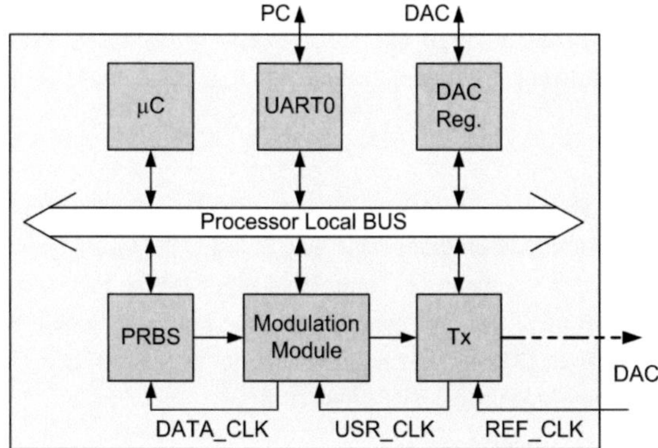

Fig. 3.3: FPGA architecture. The FPGA contains high speed modules for the generation of a pseudo random bit sequence (PRBS generation),

All components are connected to a common bus, which is referred to as the processor local bus (PLB). Through this bus the processor has access to all different components taking over general control tasks and supervision. The communication between PC and FPGA is implemented through a serial interface (universal asynchronous receiver transmitter – UART) whereas the DAC is connected through a simple register interface. The Tx-Module comprises all functionalities related to the high speed interfaces (multi gigabit transceivers – MGT) that are used to connect the FPGAs to the DACs. This includes data mapping and bit synchronization. To synchronize the high speed modules, each module provides a clock (e.g. USR_CLK, DATA_CLK in Fig. 3.3) to its preceding module. This is necessary, as the master clock for the transmitter is provided by the DACs. Since this transmitter can support multiple modulation formats, different modulation modules are available. All those modules have a standardized interface towards neighboring components to guarantee that they are easily interchangeable. The modulation module can also contain a pattern that was generated offline. A pseudo random bit stream generation (PRBS generation) according to [70] is implemented to supply the raw data for modulation modules, and this architecture is also ready for expansions like frame generation, forward error correction (FEC), or other encoding techniques. Details on the implementation can be found in Appendix A.6.

3.4 Experimental Results

3.4.1 Electrical Characterization of the DAC output

To characterize the functionalities of the FPGA design and the synchronization with the DAC, the electrical output signal of the DAC is measured.

3.4.1.1 Eye Diagram of a Two Level Signal

Here, measurements after synchronization of one FPGA with a DAC are displayed. An onboard PRBS generator has been used to compute a PRBS with a length of $2^{15}-1$ in real-time. The electrical eye diagram at the output of the DAC at 28 Gbit/s is displayed in Fig. 3.4.

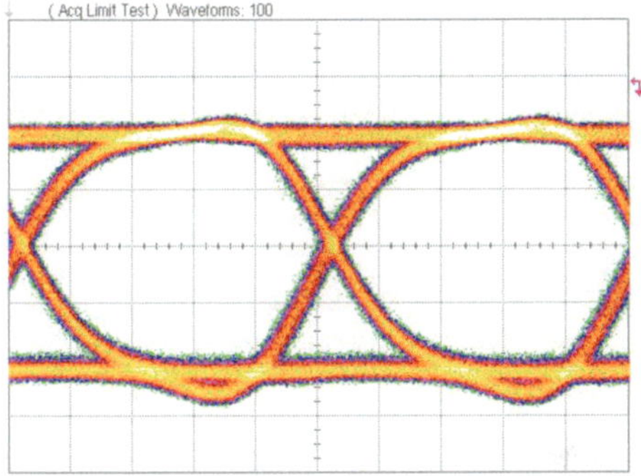

Fig. 3.4: Two level eye diagram at the output of the DAC at 28 GSa/s. The PRBS at 28 Gbit/s has a length of $2^{15}-1$ and is generated in real-time on the FPGA chip. The voltage is displayed with 150 mV/div.

All 6 bits of the DAC are set to the same PRBS resulting in an eye diagram with two levels.

3.4.1.2 Eye Diagram of a Four Level Signal

To generate a four level signal, a PRBS at 56 Gbit/s is generated (length $2^{15}-1$). After generation in real-time, the PRBS is encoded in a 4 level signal as describe in A.6.4 – also referred to as pulse amplitude modulated (PAM4). In addition, the signal is modified to compensate for the nonlinear modulator transfer function (see Fig. A.23). After the modulation module, the signal is converted to the analog domain in the DAC. The output eye diagram is shown in. Fig. 3.5.

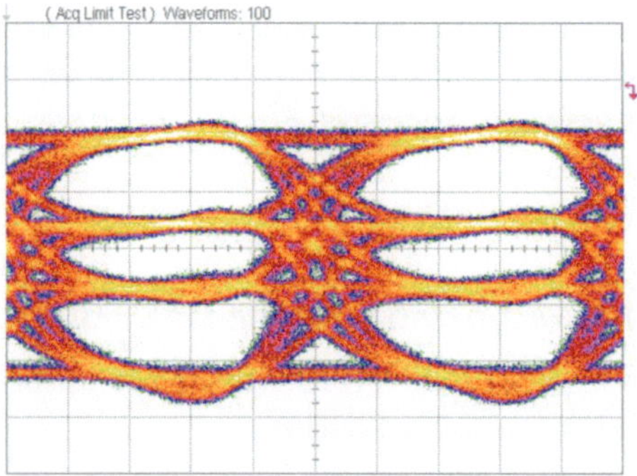

Fig. 3.5: Eye diagram of a PAM4 signal (four level) at the output of the DAC at 28 GSa/s with a PRBS at 56 Gbit/s and a PRBS length of $2^{15}-1$. The PRBS is generated in real-time on the FPGA chip. The voltage is displayed with 150 mV/div.

One can see that the pre-compensation is not as strong as illustrated in Fig. A.25. This is due to the fact that the amplitude of the drive signal is reduced compared to the theoretical maximum. This is necessary to reduce the loss of effective resolution as described in A.6.5.

3.4.2 Optical Characterization of the Transmitter

In the following section, the experimental results of the optical characterization of the transmitter are presented. This work has been published in 2010 on the conference on Signal Processing in Photonic Communications [1].

Software–defined multi–format transmitter with real–time signal processing for up to 160 Gbit/s

D. Hillerkuss, R. Schmogrow, M. Huebner, M. Winter, B. Nebendahl, J. Becker, W. Freude, and J. Leuthold

in *Signal Processing in Photonic Communications* (Optical Society of America, 2010), paper SPTuC4. [1]

3.4.2.1 Experimental Setup

The transmitter implementation and the setup for the experimental characterization is shown in Fig. 3.6. Two Xilinx Virtex 5 (XC5VFX200T) FPGAs running at a clock frequency of up to 234.38 MHz generate two PRBS (2^{15}-1) bit streams at up to 60 Gbit/s to allow testing without an external data source. The two bit streams are encoded onto 6 bit encoded multi-level signals for I and Q-channel, respectively. The 6 bit data stream is transmitted to the DACs using 24 high-speed interconnects operating at up to 7.5 Gbit/s each. The DACs generate the corresponding analog waveforms at up to 30 GSa/s. The levels of the analog waveform are optimized to compensate for the nonlinearity of the Mach-Zehnder structure in order to achieve an optimized signal constellation. The amplified analog waveforms drive the nested Mach-Zehnder IQ modulator. To generate OOK or BPSK, the output of the second DAC is switched off. For OOK, the bias voltage has to be adjusted. The optical local oscillator of the optical modulation analyzer serves as an optical carrier input for the modulator. The optical output signal of the modulator is split, delayed by 5.3 ns (106 Symbols @ 20 GBd) for decorrelating both signal paths, and afterwards multiplexed to two orthogonal polarizations for generating the polarization multiplexed (dual polarization – DP) signal. Single polarization signals are generated by disconnecting one path of the polarization multiplexing setup. The amplified output signal is received using an optical modulation analyzer Agilent N4391A as well as a 70 GHz photodiode connected to a digital communications analyzer with 70 GHz electrical bandwidth. The real-time oscilloscope (bandwidth 13 GHz) allowed us to perform experiments up to a symbol rate of 20 GBd for BPSK, QPSK and 16QAM. The optical modulation analyzer employs a raised cosine filter for pre-filtering the signal, and a subsequent finite impulse response (FIR) equalization filter with 21 taps.

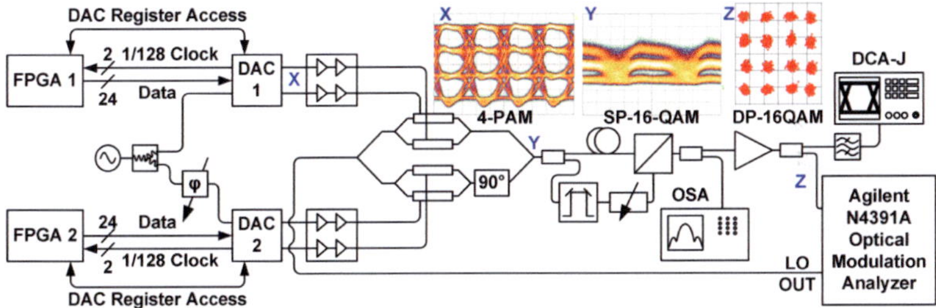

Fig. 3.6: Transmitter implementation and experimental setup. Two PRBS sequences (length 2^{15}-1) are generated by PRBS generators (PRBS Gen.) on the field programmable gate arrays FPGA1,2. The bit streams are encoded in the modulation format encoder (Modul. Encoder) generating the desired signal constellation. The GTX module uses 24 GTX-transmitters on each FPGA to feed the digital-to-analog converters DAC1,2 that generate the analog drive signal. The clock for the DACs is provided by an external synthesizer and distributed to the DACs that provide a divided clock for the FPGAs. Each FPGAs communicates with one DAC using the DAC register access. The amplified DAC output X drives the IQ modulator. The optical modulation analyzer N4391A provides a local oscillator output (LO OUT) as optical carrier to be modulated. The output Y of the modulator is split, decorrelated and polarization multiplexed, then amplified and split: The two outputs are fed to the N4391A Z and to a 70 GHz photodiode connected to a digital communications analyzer (DCA) with an electrical bandwidth of 70 GHz.

3.4.2.2 Results

Fig. 3.7 shows measured eye diagrams and constellations. A signal quality of $Q^2 = 26.8$ dB and $Q^2 = 25.9$ dB is recorded for SP-OOK at 20 Gbit/s and 30 Gbit/s, respectively. An error vector magnitude (EVM) of 5.6 % is achieved for BPSK, while the EVM for QPSK and 16QAM with polarization multiplexing is below 10 % and 7.8 %, respectively. Bit error ratio (BER) measurements were error-free for 2×10^6 symbols of DP-QPSK and a BER of 3.90×10^{-5} for 16 QAM. The power consumption of FPGAs, DACs and driver amplifiers is below 75 W.

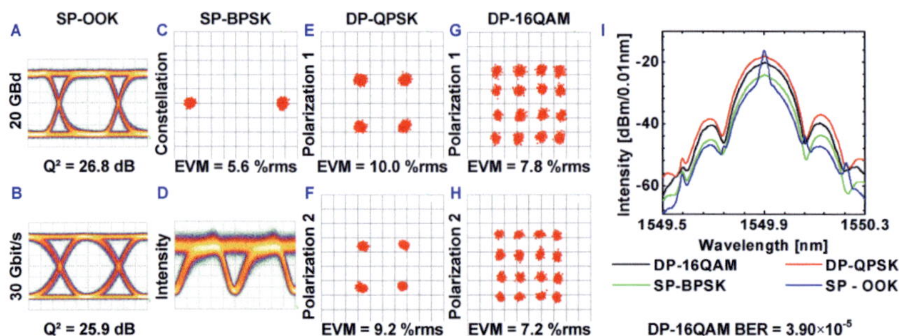

Fig. 3.7: Eye and constellation diagrams of single polarization on-off keying (SP-OOK) (A,B), single polarization binary phase-shift keying (SP-BPSK) (C,D), dual polarization quadrature phase-shift keying (DP-QPSK) (E,F) and dual polarization 16 quadrature amplitude modulation (DP-16QAM) (G,H) at 20 GBd (A, C, D, E, F, G, H,) and 30 GBd (B). Single polarization measurements are done by disconnecting one arm of the polarization multiplexing setup. The transmitter output spectra (I) are measured using a resolution bandwidth of 0.01 nm.

3.5 Summary

An optical multi-format transmitter has been developed. To this end, a feasible architecture for the implementation of an optical multi-format transmitter has been chosen consisting of field programmable gate arrays (FPGA) and digital to analog converters (DAC). The FPGA chips feature high-speed digital ports that are able to provide the required amount of data for the DAC chips running at up to 30 GSa/s. For online data generation on the FPGA a multi gigabit pseudo random bit stream generator has been implemented. The transmitter has been demonstrated to be capable of generating OOK, BPSK, QPSK and 16QAM at symbol rates up to 30 GBd. All modulation formats showed a BER well below the FEC threshold.

3.6 Outlook

To this date, the transmitter has been used for several experiments of which only a fraction has been published [3, 5-8, 13, 31, 32, 84, 85] and the developed architecture has proven to be extremely versatile as the transmitter now supports modulation formats up to 64QAM at 28 GBd [3], OFDM up to 101.5 Gbit/s [4, 31], and advanced pulse shaping methods at up to 150 Gbit/s [5, 6, 32]. The transmitter programming is still under active development and has become a standard tool in our experiments.

4 Optical Comb Generation

Advanced optical multiplexing schemes offer increased spectral efficiencies compared to standard wavelength division multiplexing (WDM) at the price of optical carriers that are stabilized with respect to each other in frequency. Examples for this are all-optical orthogonal frequency division multiplexing (OFDM) [8], no guard interval OFDM [86], coherent wavelength division multiplexing (Co-WDM) [45] and Nyquist WDM [5, 6, 13, 87]. In addition to the precise control of carrier frequencies for most formats [6, 8, 45, 86, 87], the carrier phases with respect to the symbol slot in case of coherent WDM have to be controlled [45]. Even though this could be achieved with an active stabilization of laser sources, optical frequency comb sources are ideally suited for this task as they are able to generate a large number of optical carriers with stable frequencies and a fixed carrier spacing. Several schemes have been proposed and implemented that generate these carriers using optical modulators [45, 46], recirculating frequency shifters [40, 47, 48], and mode-locked lasers (MLL) with spectral broadening in highly nonlinear fibers (HNLF) [7, 49]. In this work, optical modulators are used as a standard solution to generate a small number of optical carriers for proof of principle demonstration. The implementation of this comb generator is straight forward and can be found in literature [45]. Here, a novel scheme using a MLL and a HNLF is demonstrated that allows for the generation of a large number of optical carriers. This scheme was first introduced within the research towards this thesis [7]. It is the first scheme to generate a frequency comb with sufficient quality for communications that spans most of the central communication band. This chapter has been accepted for publication in the IEEE Photonics Journal in August 2013 [88].

High-Quality Optical Frequency Comb by Spectral Slicing of Spectra Broadened by SPM

D. Hillerkuss, T. Schellinger, M. Jordan, C. Weimann, F. Parmigiani, B. Besan, K. Weingarten, S. Ben-Ezra, B. Nebendahl, C. Koos, W. Freude, J. Leuthold

Accepted for publication in IEEE Photonics Journal, August 2013 [88].

This paper introduces a spectral slicing technique that extends the useful spectral range of frequency combs generated through self-phase modulation (SPM) of mode-locked laser pulses. When generating frequency combs by SPM, the spectral range with high-quality carriers is usually limited due to spectral minima carrying too little power. To overcome these limitations, we combine suitable slices of broadened and nonbroadened spectra. The concept was experimentally verified: A total number of 325 consecutive equidistant subcarriers span a bandwidth of 4 THz. All subcarriers have an optical carrier-power-to-noise-power-density ratio (OCNR) larger than 25.8 dB (0.1 nm) and were derived from one mode-locked laser with a mode linewidth of approximately 1 kHz. The signal quality of the comb and in particular of each subcarrier was ultimately tested in a terabit-per-second communication experiment. The comb quality allowed us to transmit 32.5 Tb/s over 225 km with 100 Gb/s dual polarization 16-ary quadrature amplitude modulation (16QAM) signals on each of the subcarriers.

4.1 Introduction

Optical frequency combs that offer hundreds of carriers of highest quality across a large spectral band have drawn a tremendous amount of attention within recent years. Research on frequency combs culminated in the Nobel Price that was awarded to John Hall and Theodor Hänsch in 2005 for their work in laser-based precision spectroscopy [89, 90]. Optical frequency combs were recognized as a useful tool not only for spectroscopy but also for various other applications such as optical and microwave waveform generation [91, 92], optical signal processing [93, 94], fiber-optic communication [7, 8, 12, 40, 45-49, 95, 96], optical coherence tomography [97] and precision ranging [98]. Depending on the application, the individual linewidth, the line spacing or the total number of lines (e. g., spanning an octave for self-referencing) are important and ultimately influence the design.

Optical frequency combs are most attractive for the latest generation of optical multi-carrier communication systems, where information is encoded in so called superchannels. To generate Tbit/s superchannels, data is encoded on multiple closely spaced, equidistant optical carriers and subsequently multiplexed. Such systems require a precisely controlled carrier spacing, which is inherently provided by optical comb sources. Here, a single comb source could replace hundreds of lasers, which would otherwise have to be controlled precisely in their relative and absolute frequencies. Prominent examples of such superchannel communication systems are based on all-optical orthogonal frequency division multiplexing (OFDM) [8], no guard interval OFDM [86], coherent wavelength division multiplexing (Co-WDM) [45], and Nyquist WDM [5, 6, 12, 87]. However, such schemes require a precise control of carrier frequencies [6, 8, 45, 86, 87], and in rare cases even of the carrier phases with respect to the symbol time slot [45]. Fortunately, in optical frequency comb sources, the carrier spacing is extremely precise, and strictly linear time dependencies of the phases between carriers are maintained, making them ideally suited for this application.

However, in order to use the carriers of an optical frequency comb for communications they must have a sufficiently narrow spacing (e.g. 12.5 or 25 GHz) and offer sufficient quality. This actually means that

- All the optical carriers should be of equal power (or follow a defined spectral distribution). This is needed to achieve a similar performance of the data encoded on the carriers and to avoid one modulated carrier limiting the performance of all neighbors through crosstalk. Equal power in all carriers typically can be achieved through equalization of a generated frequency comb [8, 49].

- The carriers must have a minimum carrier-power to noise-power-density ratio (CNR). The CNR ultimately limits the amount of information that can be encoded on a single subcarrier. The CNR directly defines the maximum achievable signal-to-noise ratio (SNR), for which minimum values can be found in tables.

- The linewidth of all carriers has to be narrow. A narrow linewidth (low phase noise) is fundamentally needed - particularly for transmission systems making use of phase encoded signals. Coherent transmission systems typically need a laser linewidth that is significantly smaller than 100 kHz.

Quite a few schemes have been experimentally investigated for use in optical communication systems that can fulfill these requirements to a smaller or larger extent. Among these are comb generators based on optical modulators [45, 46, 99], also in the form of recirculating frequency shifters [40, 47, 48], micro resonators [95, 100, 101], and mode-locked lasers (MLL) with spectral broadening in highly nonlinear fibers (HNLF) [7, 49]. All these schemes have different advantages and drawbacks: Schemes using modulators typically have a sufficient carrier quality but only offer a few lines and the operation points of the modulators have to be controlled precisely. Schemes that exploit self-phase modulation (SPM) typically provide a sufficient number of subcarriers, but the CNR is low at the spectral minima and at the outer edges of the spectrum.

In this paper, we present a comb generator scheme based on slicing of spectra broadened in highly nonlinear fibers. The scheme extends the usable bandwidth of frequency combs generated through SPM by removing the minima [2] in broadened spectra. The scheme promises a large number of carriers with ample power and CNR for optical transmission systems by spectrally composing suitable subspectra [8]. In our demonstration, we generate a total number of 325 carriers, which we derived from a single MLL. The MLL was measured to provide a linewidth of approximately 1 kHz. The CNR of all carriers of the frequency comb was larger than 25.8 dB and the quality was proven in a transmission experiment.

4.2 Spectral Minima in Self–Phase Modulation

Spectral broadening through self-phase modulation (SPM) can generate extremely broad frequency combs. However, as already observed by Stolen et al. in 1978 [2], spectral minima occur for phase shifts significantly larger than π. Typically, these minima are explained as destructive interference of spectral components that have experienced a relative phase shift of π [102].

In a simplified picture, the generation of the spectral minima in SPM can be illustrated as shown in Fig. 4.1. The intensity of an optical impulse is displayed in a moving time frame $z - v_g t$ (group velocity v_g) along the propagation direction z, Fig. 4.1 (a). Due to the Kerr nonlinearity, the optical phase shift $\varphi \propto I$ is in proportion to the local intensity I. The slopes of the impulse lead to a frequency offset Δf, Fig. 4.1 (b). If the maximum phase shift φ_{max} is larger than π, then above (and also below) $\Delta f = 0$ there will be pairs of points with the same Δf but a relative phase shift $\Delta\varphi = \pi$. In this simple picture, this will lead to destructive interference of these spectral components, and therefore to minima in the spectrum. In our example, these frequencies are labeled Δf_1 and Δf_2.

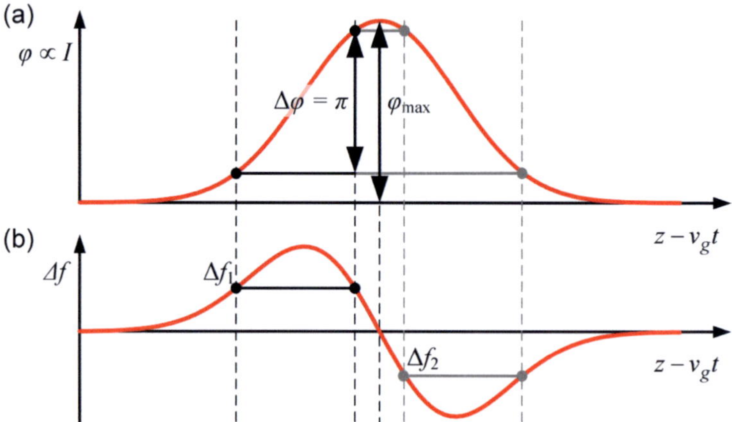

Fig. 4.1: Schematic phase and frequency shift in self-phase modulation of a single impulse with intensity I in a moving time frame $z - v_g t$ (propagation coordinate z, time t, group velocity v_g). (a) Nonlinear phase shift $\varphi \propto I$. The phase shift has a maximum φ_{max} at the peak of the pulse and translates to a frequency shift Δf as seen in (b). In a simplified picture, each frequency shift occurs twice within one pulse (see example in this figure), but with a different phase. If the maximum phase shift φ_{max} is larger than π, there will be at least two pairs of points (Δf_1 and Δf_2), one with a positive Δf_1 and one with a negative frequency shift Δf_2, which have a phase difference of $\Delta\varphi = \pi$. This phase shift then leads to spectral destructive interference at these frequency shifts Δf_1 and Δf_2. In our example these frequency shifts are symmetric around $\Delta f = 0$, which is not necessarily the case for asymmetric impulses or in the presence of dispersion.

Fig. 4.2 presents simulated and measured power spectra that exhibit these minima. The simulated example spectra Fig. 4.2 (a) are computed for a single spectrally broadened pulse. If $\Delta\varphi \geq \pi$, spectral minima develop. As seen in the figure, the number of minima m allows for an estimation of the maximum phase shift φ_{max} according to

$$\varphi_{max} \approx \frac{2m+1}{2}\pi \, . \tag{4.1}$$

This equation is related to the equation (4.1.14) in [102], which gives a similar relation for the number of spectral maxima resulting from a certain maximum phase shift.

The measurement in Fig. 4.2 (b) relates to the mode locked laser (MLL) with a repetition rate $R = 12.5\,\text{GHz}$, which was used for all experiments in this paper. Its comb line spectrum was amplified to the specified average powers and broadened by SPM. The minima are marked with white symbols "v". Because interference is involved, any instability in the optical MLL power leads to a frequency shift of these minima, and therefore to a pronounced instability of the neighboring comb lines.

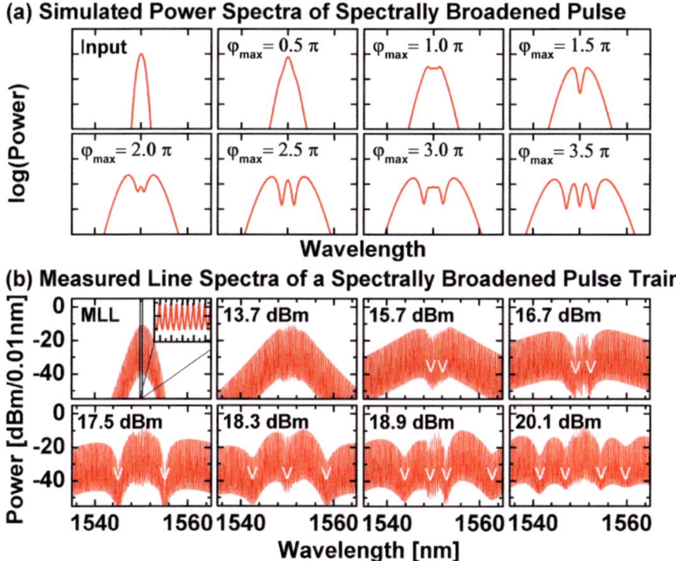

Fig. 4.2: Simulated and measured spectra for spectral broadening through self-phase modulation in a highly nonlinear fiber. (a) Single pulse with its spectrum labeled "input". After spectral broadening in a highly nonlinear fiber (HNLF), the maximum nonlinear phase shift is φ_{max}. It increases proportional to the input power, which increases from left to right and from upper to lower row. The spectra exhibit minima for maximum nonlinear phase shifts significantly larger than π [2]. (b) Non-broadened line spectrum of the mode locked laser (MLL), and line spectra broadened by SPM in the HNLF. The average input power to the HNLF is given in dBm. Spectral minima are marked with a white "v". For higher powers, not all minima are within the displayed bandwidth of 30 nm. The spectra were measured at the output of the HNLF in the setup seen in Fig. 4.4.

4.3 Comb Generation by Spectral Slicing

The following scheme generates a frequency comb with a large bandwidth and a high carrier quality for all carriers. We propose to compose this high quality frequency comb from a seed spectrum and one or more broadened spectra [8]. We replace carriers at spectral minima of broadened spectra with a low carrier power and CNR by carriers from differently broadened spectra with a high carrier quality. This process has the four steps shown in Fig. 4.3. First, we generate the seed spectrum using an optical pulse source, such as a MLL. From this seed spectrum, we then generate N optical spectra via spectral broadening elements (NLE), such as a highly nonlinear fiber (HNLF). Next, we use spectral slicing and composing in a reconfigurable optical multiplexer (e.g. waveshaper). Here, the usable parts of the seed spectrum and the N additional spectra are sliced and combined to form a continuous broad output spectrum. This effectively replaces all unstable and low-power parts of the optical spectra. Finally, the spectrum is equalized. The bottom inset in Fig. 4.3 illustrates this process for $N = 1$.

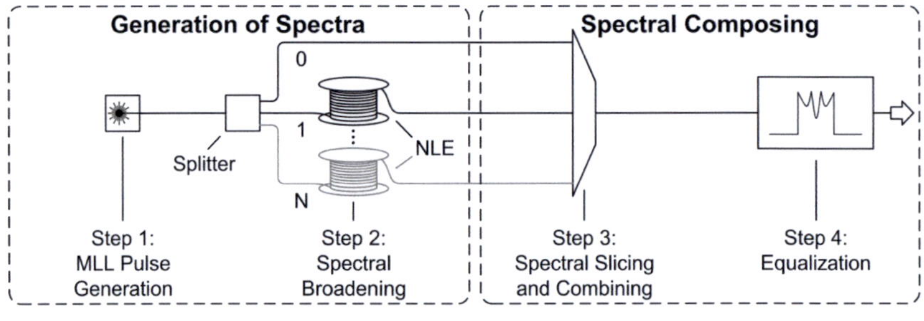

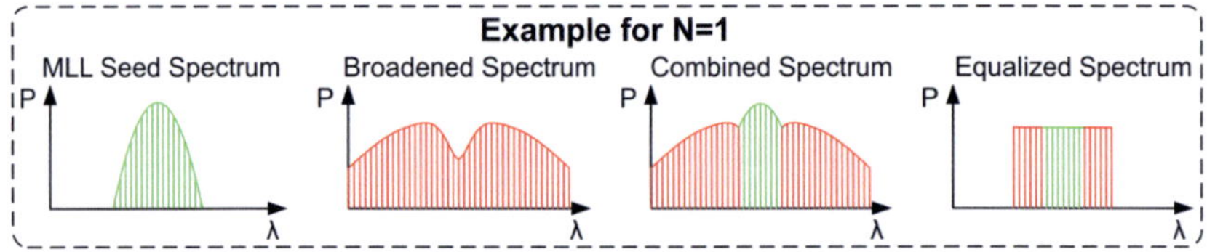

Fig. 4.3: Concept of the optical comb generator by spectral slicing. First, N optical spectra are generated; a mode locked laser (MLL) generates the initial seed spectrum (as shown in green) that is spectrally broadened in a nonlinear element (NLE) to generate N broadened spectra with a large number of carriers (as shown in red). Second, the output spectrum is generated from the N spectra using spectral slicing and composing; for this, the generated spectra are sliced and combined to form a wide spectrum. This wide spectrum is subsequently equalized to obtain the output spectrum of the comb generator. The lower part in this figure shows exemplary spectra for two combined optical spectra.

For current implementations of this scheme with discrete components, the following points have to be kept in mind as they could limit the overall performance and/or usability for different applications:

- A poor filter-stop band could limit the performance in several ways.

  o As a result of a poor filter-stop band multipath interference between carriers from different paths 0, 1, … N in Fig. 4.3 could lead to low frequency fluctuations in the kHz range due to thermal path length changes. This could result in phase and amplitude fluctuations of the comb lines. This did not pose an issue in our experiments as the 40 dB filter extinction ratio of the waveshaper was sufficient as outlined in section 4.

- o The filter's extinction ratio often is not sufficient at the edge of the filters when going from stop-band to pass band. In our experiment we indeed noticed some small crosstalk in the two lines where spectral slices are merged. However, this multipath interference should not be mistaken for coherent crosstalk in communication systems [103], as the interfering laser lines do not carry any data at that point. And indeed, in our case it did not lead to any signal degradation.

- o A poor filter-stop band could also degrade the overall CNR as a result of summing up noise floors from different paths. This however is quite unlikely, as the noise floor of the interfering spectrum would have to be much higher than the noise floor in the desired spectrum as it is attenuated by the filter extinction. In our experiment this was most definitely not an issue as we started with a very good CNR that was further suppressed in the filters.

- As the different paths in the comb generator consist of different lengths of fiber, a decorrelation of the different segments of the comb could occur if the lengths are equal to or larger than the coherence length. This could lead to very small frequency shifts between spectral comb slices if the center frequency of the source laser drifts. Due to the short fiber length of 100 m and a coherence length in the order of 100 km, this was no issue in our experiments.

Of course, in an ultimate integrated solution, the best stability could be by achieved by miniaturization that includes on chip nonlinear elements instead of HNLFs.

4.4 Experimental Setup of the Broadband Spectrally Sliced Comb Source

Fig. 4.4 shows our experimental setup. A passively mode locked laser (MLL – Ergo XG) with a repetition rate of 12.5 GHz and a pulse width below 2 ps generates the initial seed spectrum (Fig. 4.4a). The spacing of the spectral lines equals the repetition rate, while the pulse width and shape determine the spectral envelope. The chirp of the MLL is adjusted by 5.4 m of standard dispersion compensating fiber (DCF). Different fiber lengths were tested, and the length generating the largest amount of SPM at the output of the HNLF was chosen for the experiment. The spectrum is amplified, and a 5 nm bandpass filter suppresses the amplifier noise. The laser signal is now split in two parts. One part is spectrally broadened by SPM (see Fig. 4.4b) in a highly nonlinear photonic crystal fiber (HNLF). The HNLF is a photonic crystal fiber with the following parameters: length 100 m, dispersion 1.25 ps/nm/km, mode field diameter 2.8 ± 0.5 µm, attenuation <9 dB/km, nonlinear coefficient 19 W^{-1} km^{-1}. The other part is not broadened; instead, it is just passed through. Finally, the waveshaper slices, combines, and equalizes the broadened and the original MLL spectrum, thus producing the final output spectrum, Fig. 4.4 (c). To show the suppression and estimate possible fluctuations of the power in individual lines, we have performed the measurement shown in Fig. 4.4 (d). The most critical carriers are the carriers at the edges, where we only achieve a suppression of > 16 dB. All other carriers have a suppression > 30 dB and typically 40 dB. This corresponds to a power fluctuation of < 3 dB for the carriers at the edges and a fluctuation of < 0.6 and typically <0.2 dB for the central carriers. The speed of these fluctuations only depends on linewidth and frequency stability of the MLL and the stability of the interferometer. In our experiment, these fluctuations were extremely slow due to the high stability of the MLL.

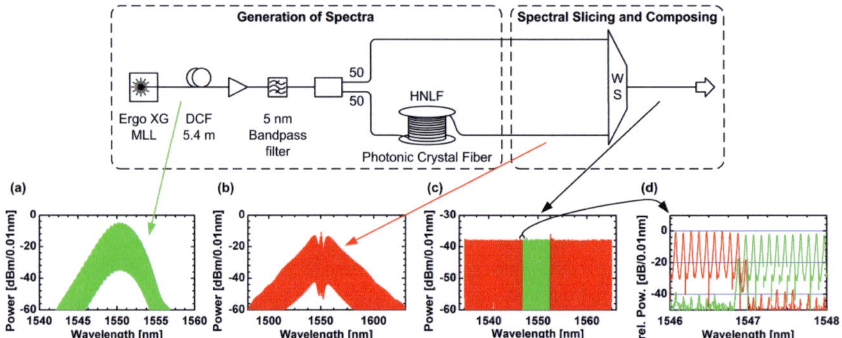

Fig. 4.4: Experimental setup of the ultra-broad spectrally sliced comb. The output spectrum of the mode locked laser (MLL – Ergo-XG, repetition rate 12.5 GHz) is displayed in the first inset (a – shown in green). The MLL pulses are amplified and filtered in a 5 nm bandpass filter. The amplified pulses are split, and one path is passed through and the other path is spectrally broadened in a highly nonlinear photonic crystal fiber (b – shown in red – HNLF). A waveshaper (WS) performs the spectral slicing and composing to generate the output spectrum (c – shown in green and red). The suppression of the neighboring spectra at one of the edges is shown (d). The suppression at the carriers next to the edges is > 16 dB. For all other carriers, a suppression >30 dB, typically 40 dB is achieved.

The resulting flat frequency comb shown in Fig. 4.4(c) has 325 high-quality carriers that have been used for multiple Terabit/s transmission experiments [8, 12] with aggregate data rates of up to 32.5 Tbit/s [12]. In these experiments, we carefully measured data transmission on each and every spectral line of the generated frequency comb. We observed a similar performance for all subcarriers and did not see any impact of multipath interference [8, 12].

4.5 Characterization

4.5.1 Linewidth Characterization of the Mode Locked Laser

The linewidth of the individual modes of the MLL fundamentally limits the linewidth of the carriers in the broadened comb. We therefore measured the linewidths of five modes of the MLL, which we found to be approximately 1 kHz. We were limited to these five modes around 1550 nm due to the fixed wavelength of the narrow linewidth reference laser.

The linewidth of lasers is often measured using a delayed self-heterodyne technique [104, 105], in which the laser is split and one copy is decorrelated in a fiber delay line before being mixed with the original laser line. However, from our transmission experiments with this comb source [12], we expect an extremely low linewidth. This would results in an excessively long fiber length required for decorrelation [104]. Such a long fiber delay renders self-heterodyne measurements extremely susceptible to acoustic noise and other environmental influences. As such types of noise have significant frequency components in the same order of magnitude as the linewidth of interest, self-heterodyne measurements were not practical.

We therefore chose to measure the linewidth through a heterodyne measurement [105]. Normally, one would perform the measurement by combining the laser line under test with a narrow linewidth laser (NLWL) and then detecting the signal in a photodiode. The photocurrent is then analyzed in an RF-spectrum analyzer. However, due to the low measurement speed of these analyzers, this scheme requires wavelength tracking to compensate for slow drifts of the laser wavelengths [105]. Therefore, we decided for a new technique. To measure the linewidth of the laser without the drift of the laser wavelengths we chose to perform the measurement using real-time acquisition and subsequent spectral analysis of the recorded signal, see Fig. 4.5(c). In this scheme we used a coherent receiver (Agilent optical modulation analyzer) with a narrow linewidth laser (NLWL) as external local oscillator (LO). The receiver has an electrical bandwidth of 32 GHz per inphase and per quadrature channel and a sampling rate of 80 GSa/s per channel. The NLWL had a linewidth of approximately 1 kHz, and the down-converted optical pulse train of the MLL was recorded for a time span of 12.5 ms. The advantage of this new technique is that we can restrict the data acquisition to a time interval within which the relative laser frequency drift is significantly smaller than the linewidth. This way there is no need for performing active wavelength tracking even though the frequency might drift over several kHz within a few seconds. This slow frequency drift, however, is not critical for optical communication systems, as the frequency offset compensation in coherent receivers can easily compensate for it.

First, we analyze a subset of 2^{28} samples (3.35 ms). Within this short time window, the influence of frequency drift of the two lasers is marginal. After performing a fast Fourier transform (FFT) on the temporal data, we analyze each of the 5 MLL lines within the receiver bandwidth of 64 GHz centered on 1550 nm. To this end, we process spectra with a width of 2^{15} samples (~9.8 MHz) centered at each MLL line. All spectral lines look alike, so that we only picture one line in Fig. 4.5. The full-width half-maximum (FWHM) of the power spectra of all five analyzed lines is found to be 1.9 kHz. The FWHM of a Lorentzian fit, red lines in Fig. 4.5(a,b,d), is 2.2 kHz. Because the side lobes at ± 590 kHz were included in the fitting process and because of the limitation in the sampling duration and the 1 kHz linewidth of the reference laser, the inferred line width is a worst-case estimate of the actual 3 dB bandwidth.

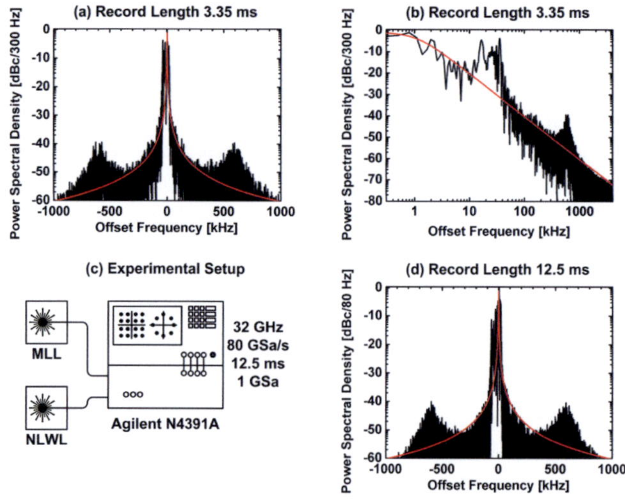

Fig. 4.5: Linewidth characterization of the mode locked laser (MLL). As shown in the bottom left subfigure (c), the mode locked laser (MLL) is recorded for 12.5 ms in a coherent receiver. The local oscillator is a narrow linewidth laser (NLWL) with a linewidth of approximately 1 kHz. The upper two subfigures (a,b) display the power spectral density (shown in black) of one of the MLL lines mixed with the NLWL in an observation interval of 3.35 ms. The full width half maximum (FWHM) of the measured spectrum is less than 2 kHz, while the FWHM of the Lorentzian fit (shown in red) is 2.2 kHz. Side peaks at ± 590 kHz are visible in addition to the Lorentzian line shape. The frequency drift dominates the spectrum for a longer record length of 12.5 ms (d). As reference, we also display the Lorentzian fit obtained from the shorter record length.

Next, we process the full dataset. In this case, the spectra are distorted by the drift of the two lasers, Fig. 4.5(d). For the recorded 5 MLL lines we observe linewidths between 260 Hz and 1.33 kHz. From these measurements, we conclude that the characterized modes of the MLL have a linewidth of approximately 1 kHz. The side lobes to be seen in Fig. 4.5 were also observed when investigating the MLL spectra with self-heterodyne measurements. However, due to environmental fluctuations, this measurement was not suitable for determining the linewidth.

4.6 CNR and OCNR Characterization of the Comb Source

The signal-to-noise ratio (SNR) and therefore the minimum carrier-power to noise-power-density ratio (CNR) have to be as as large as possible in order to allow for reliable transmission of information.

We first implemented the comb source to study advanced multiplexing schemes at aggregate data rates beyond 10 Tbit/s [8, 12]. When investigating the noise characteristics in these experiments, we observed circular noise "clouds" around all constellation points. A statistical analysis of the error vector pointing from a nominal constellation point to the momentarily received signal yields the distribution of the noise in the system. From our statistical analysis in [12], we inferred a Gaussian distribution of the corresponding noise [12]. We conclude from these observations that the limiting factor for the signal quality is additive Gaussian noise due to amplified spontaneous emission in the optical amplifiers in the system, and not the phase noise of the optical carriers themselves. In the presence of significant phase noise the distribution of the received constellation points would have been distorted in an angular direction.

Next, we measured the CNR of the optical comb source. The CNR is defined as the ratio of the carrier-power P_C and the noise-power-density N_0 at the position of the carrier,

$$\text{CNR [dBHz]} = 10\log_{10}\left(\frac{P_C}{N_0\Delta f}\right). \tag{4.2}$$

Here, the noise power density N_0 is normalized to a bandwidth of $\Delta f = 1$ Hz.

Characterization setup, measurement principle, and results of the CNR measurement are shown in Fig. 4.6. To measure the noise floor between the carriers with a spacing of 12.5 GHz, we used a high resolution optical spectrum analyzer (Apex AP2050A). The comb generated by spectral slicing has been implemented as described in Section 4.4 and Fig. 4.4. The waveshaper (WS) now serves a different purpose. It is programmed to realize a bandpass filter function with a 3 dB bandwidth of 60 GHz. The center of the bandpass coincides with the carrier of interest. The filtered spectrum is then amplified in a low noise EDFA (not depicted in Fig. 4.4, noise figure NF = 3.7 dB) to increase the power of the measured noise to a value significantly larger than the receiver noise of the spectrum analyzer. To ensure that the measurement is not limited by the electronic noise of the spectrum analyzer, the spectra are measured with three different resolution bandwidths of 50 pm, 20 pm, and 10 pm. The noise level at the carrier frequency is determined by linear interpolation between the noise levels on both sides of the carrier, see illustration in Fig. 4.6(b). All noise density measurements are normalized to a reference bandwidth of 1 Hz. Noise density measurements with different resolution bandwidths deviated by less than 0.6 dB, therefore electronic noise had only minimal influence on the measurement. Both, the amplified seed and the broadened spectrum, were present at the waveshaper inputs simultaneously during the CNR measurements.

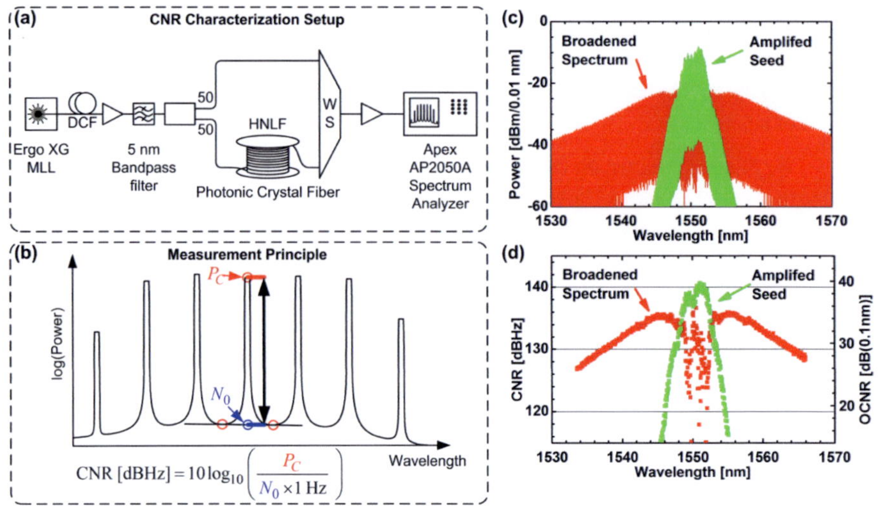

Fig. 4.6: Carrier-power to noise-power-density ratio (CNR) characterization of the comb source. (a) For the characterization of the CNR, the generation of spectra is implemented as illustrated in Fig. 4.4. Instead of equalization and spectral slicing, the waveshaper (WS) extracts 60 GHz out of the generated spectra (see top right) around the carrier to be characterized. A low noise EDFA amplifies the excerpt for characterization in the Apex AP2050A high resolution spectrum analyzer. The CNR is measured as illustrated in (b). The peak of the carrier is measured and the noise is determined by pseudo linear interpolation between the minima between the measured carrier and its neighbors. The obtained CNR values are shown in the plot (bottom right).

In optical communication systems, a commonly used measure for communication signals is the optical signal-to-noise ratio (OSNR), which is given as the signal power in relation to the noise power density normalized to a bandwidth of 0.1 nm. This measure was established for historic reasons as 0.1 nm used to be the minimum resolution of optical spectrum analyzers. We therefore also provide the OCNR

$$\mathrm{OCNR\,[dB(0.1nm)]} = 10\log_{10}\left(\frac{P_C}{N_0}\frac{\lambda_c^2}{c\,\Delta\lambda_{\mathrm{ref}}}\right), \tag{4.3}$$

with the speed of light c. Here, the noise-power density is normalized to a bandwidth of $\Delta\lambda_{\mathrm{ref}} = 0.1$ nm at a reference wavelength $\lambda_c = 1550$ nm.

The CNR (and OCNR values) of the amplified seed spectrum, see green range in Fig. 4.6(d), is larger than 133.5 dBHz – reference bandwidth 1 Hz (32.5 dB(0.1nm) – reference bandwidth 0.1 nm) in the center region of the flattened comb – the green part in Fig. 4.4 (d). It varies between 114 dBHz (13 dB(0.1nm)) and 140.6 dBHz (39.6 dB(0.1nm)) over the measured bandwidth of 9.5 nm – here, we measured the CNR also measured across the spectral lines that were removed by the filter. The CNR of the broadened spectrum – the red part that was used to generate the spectrum in Fig. 4.4 (d) and without the central green part – varies between 126.8 dBHz (25.8 dB(0.1nm)) and 136 dBHz (35 dB(0.1nm)). The CNR of the broadened spectrum that was replaced by the amplified seed spectrum was measured as well.

We found that the CNR in this region was heavily fluctuating as one would expect from our discussion in Section 2. As a result we found scattered red dots as seen in Fig. 4.6d. It should be mentioned that we sliced the spectrum at the points, where the carriers of the amplified seed and the broadened spectrum had a similar CNR and also a similar power. The plots in Figure 6c and Figure 6d also show that there was a strong correlation between the CNR and the power of the individual lines.

One might argue that this measurement scheme might underestimate the CNR for real operating condition, as the central part of the comb is attenuated by up to 25 dB and noise crosstalk from the inner part of the broadened spectrum might occur. This is not the case. The input spectrum to the HNLF has the same power as the input to the waveshaper for the center part of the spectrum. Therefore, the noise levels in between the lines for both spectra are similar. This crosstalk noise is at least 20 dB below the noise of the attenuated original spectrum due to the additional insertion loss of the HNLF (>5 dB) and the extinction of the waveshaper (>40 dB).

To put these numbers in relation, we provide the required SNR for transmission of two of the most common modulation formats for current and next generation communication systems. The SNR needed to transmit a polarization multiplexed 12.5 GBd QPSK or 16QAM signal with a bit error ratio of 2×10^{-3} is 10 or 16 dB, respectively [64]. The maximum symbol rate for our frequency comb with a spacing of 12.5 GHz is 12.5 GBd at Nyquist channel spacing [12]. For this symbol rate, the SNR is approximately equal to the OCNR. Thus, our frequency comb provides a carrier quality that is sufficient for an SNR larger than 25.8 dB for all carriers.

4.7 Conclusion

We present a scheme for the generation of broad frequency combs via self-phase modulation. Differently broadened spectra are sliced and combined to a compound spectrum. Unstable, low-power spectral portions with a low CNR are removed and replaced with stable parts of the differently broadened spectra. A mode locked laser with a mode-linewidth of approximately 1 kHz serves as a seed spectrum. The generated frequency comb has 325 optical carriers and covers a bandwidth larger than 4 THz. We measured carrier-power to noise-power-density ratios (CNR) between 118.6 dBHz and 104.6 dBHz with 1 Hz reference bandwidth (39.6 dB(0.1nm) and 25.9 dB(0.1nm) with 0.1 nm reference bandwidth) for the regions of interest. The carrier quality suffices for transmission of multiple terabit/s using 16QAM [8, 12] with aggregate data rates of up to 32.5 Tbit/s [12].

5 Advanced Optical Multiplexing

OFDM and All–Optical OFDM

This chapter has been published in optics express [10]. Small corrections in Fig. 5.3, where a phaseshifter had been placed wrongly and in section 5.1.2.4, where X_2 had to be replaced by X_4 have been included in this reprint. In addition, most of the symbols have been modified to be in line with the rest of this thesis. However, the symbol X_m for the data encoded on the m-th subcarrier of an OFDM symbol and the m-th frequency sample of a signal in the frequency domain is used instead of c_{ik}, which is used in the rest of the thesis.

Simple all–optical FFT scheme enabling Tbit/s real–time signal processing

D. Hillerkuss, M. Winter, M. Teschke, A. Marculescu, J. Li, G. Sigurdsson, K. Worms, S. Ben Ezra, N. Narkiss, W. Freude, and J. Leuthold

Optics Express **18**(9), 9324-9340 (2010) [10].

5.1 The Optical Fast Fourier Transform

In [9, 10], a practical scheme to perform the fast Fourier transform in the optical domain is introduced. Optical real-time FFT signal processing at speeds far beyond the limits of electronic digital processing and with negligible energy consumption is demonstrated. To validate the method we demonstrate an optical 400 Gbit/s OFDM receiver. It performs an optical FFT in real-time on the consolidated OFDM data stream, thereby demultiplexing the signal into lower bit rate subcarrier tributaries, which can then be processed electronically.

5.1.1 Introduction

The fast Fourier transform (FFT) is a universal mathematical tool for almost any technical field. In practice it either relates space and spatial frequencies (spatial FFT) or time and temporal frequency (temporal FFT). While the former can most easily be implemented in the optical domain by means of a lens, the time-to-frequency conversion in the optical domain is more intricate. Yet, it is exactly this conversion that is needed for next generation signal processing such as required for OFDM in optical communications. The importance of the optical FFT for the implementation of next generation processors has been recognized in the past, and direct implementations of the FFT in integrated optics technology have been suggested by Marhic [106] and others. However, to this point these schemes are difficult to implement and stabilize, and they do not scale well with increasing FFT order.

In this paper, we introduce a new and practical implementation of the optical FFT. We discuss the feasibility, tolerances towards further simplifications, practical implementations, and we demonstrate the potential of the method by an exemplary implementation for a next generation OFDM system. We show that a single 400 Gbit/s OFDM channel can be demultiplexed into its constituting subchannels by optical means. The method is based on passive optical components only and thus basically provides processing without any power consumption. The scheme therefore is in support of a new paradigm, where all-optical and electronic processing synergistically interact providing their respective strengths. All-optical methods allow processing at highest speed with little – if no power consumption, and electronic processing performs the fine granular processing at medium to low bit rates.

In Section 2 we introduce the new optical FFT. In Section 3 we show how the optical FFT and IFFT can be applied to OFDM transmission systems in order to enable processing of OFDM-channels at very high bit rates without the limitations of the electronic circuits. In Section 4, we will present the results of an experimental implementation of the optical FFT for demultiplexing a consolidated OFDM signal into its OFDM subchannels.

5.1.2 The optical FFT/IFFT

In this section we introduce the new optical FFT method. We first recapitulate conventional means for performing the optical FFT and differentiate it from the commonplace FFT as done with a computer. We then show how subtle re-ordering of the various FFT circuit elements leads to a significant reduction of the complexity. It also leads to a simple and practical optical circuit with an equivalent output. Afterwards we show how this circuit can be further simplified at the cost of a small interchannel crosstalk penalty by replacing one or more stages of the FFT circuit by standard optical (tunable) filters.

5.1.2.1 Background

The fast Fourier transform (FFT) is an efficient method to calculate the discrete Fourier transform (DFT) for a number of time samples N, where $N = 2^p$ with p being an integer. The N-point DFT is given as

$$X_m = \sum_{n=0}^{N-1} \exp\left[-j2\pi \frac{mn}{N}\right] x_n \ , \quad m = 0,..,N-1 \tag{5.1}$$

transforming the N inputs x_n into N outputs X_m. If the x_n represent a time-series of equidistant signal samples of signal $x(t)$ over a time period T_s, as shown in Fig. 5.1(a), then the X_m will be the unique complex spectral components of signal x repeated with period T_s [71]. The FFT typically "decimates" a DFT of size N into two interleaved DFTs of size $N/2$ in a number of recursive stages [72] so that

$$X_m = \begin{cases} E_m + \exp\left[-j\dfrac{2\pi}{N}m\right] O_m & \text{if } m < \dfrac{N}{2} \\[2em] E_{m-N/2} - \exp\left[-j\dfrac{2\pi}{N}\left(m-\dfrac{N}{2}\right)\right] O_{m-N/2} & \text{if } m \geq \dfrac{N}{2} \end{cases} \tag{5.2}$$

The quantities E_m and O_m are the *even* and *odd* DFT of size $N/2$ for even and odd inputs x_{2l} and x_{2l+1} ($l = 0,1,2,...N/2-1$), respectively. Fig. 5.1(b) shows the direct implementation of the

FFT for $N = 4$ using time electrical sampling and signal processing. Marhic [106] and Siegman [107, 108] have shown independently a possible implementation of an optical circuit which performs an FFT. Such an implementation for $N = 4$ is shown in Fig. 5.1(c). When using an optical circuit to calculate the FFT the outputs X_m appear instantaneously for any given input combination x_n. Thus, in order to obtain the spectral components of a time series, the N time samples in interval T_s must be fed simultaneously into the circuit. This can be achieved using optical time delays as a serial-to-parallel (S/P) converter, as shown in Fig. 5.1(c).

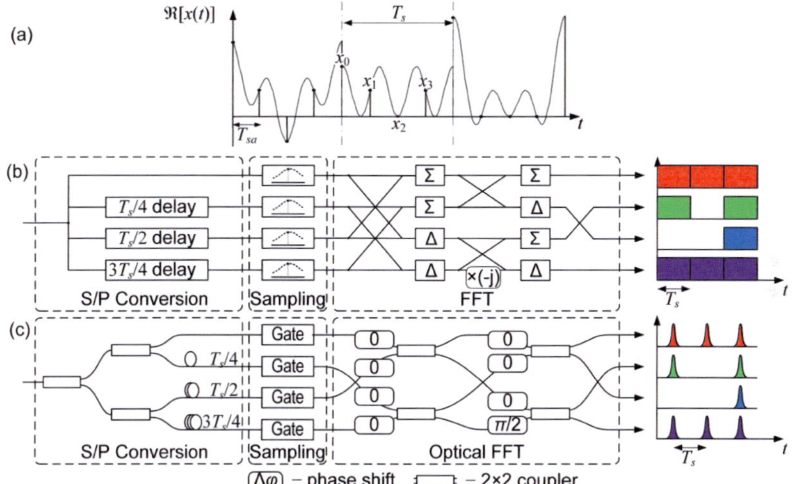

Fig. 5.1. Four point example of the traditional fast Fourier transform and its optical equivalent. (a) Exemplary signal in time sampled at $N=4$ points; (b) the structure consists of a serial-to-parallel (S/P) conversion that generates parallel samples of the signal, a sampling stage to generate the time samples x_n and a conventional FFT stage that calculates the fast Fourier transform of the sampled signal; (c) the optical equivalent of the circuit uses passive splitters and optical time delays for serial-to-parallel conversion; optical gates perform the sampling of the optical waveform; afterwards the optical FFT is computed using optical 2×2 couplers and phase shifts as described in [106]. Right-hand sides of (b) and (c) show typical output signals for input signal (a).

The optical FFT (OFFT) differs from its electronic counterpart by its continuous mode of operation. In an electronic implementation optimized for highest throughput, the optical signal is sampled and the FFT is computed from all samples x_n. Afterwards, the next N samples are taken. In the optical domain, the FFT is computed continuously. Yet, the calculation is correct only when feeder lines 1 to N contain the time samples from within their respective interval. Sampling must therefore be performed in synchronization with the symbol over a duration of T_s/N. These samples subsequently can be processed in the optical FFT stage. However, it must be emphasized that proper calculation is only possible if all samples are forwarded from one stage to the next stage in synchronism. Care must therefore be taken, that not only all waveguides interconnecting the couplers have equal delay but also maintain proper phase relations, as indicated in optical FFT stage of Fig. 5.1(c).

The optical FFT has several advantages over its electronic counterpart. First, the all-optical FFT may be used at highest speeds where electronics cannot be used. This is due to the fact that the optical sampling window sizes (e. g., with electro-absorption modulators, EAM) can be significantly shorter than electronic sampling windows of analog-to-digital converters (ADC) [109]. Additionally, since all components used in the OFFT are passive (except for tuning circuitry and time gating), the power consumption is inherently low and barely increases with complexity or sampling rate. For an exemplary 8-point FFT of a 28 GBd OFDM signal, we estimate the power consumption for the optical and the electrical sampling as follows. The power requirement for the optical sampling of 8 tributaries is dominated by the EAM driver amplifiers and would be about 14 W. In addition, several watts will be required to compensate for insertion and modulation loss of the optical gates using optical amplifiers. In comparison to this, the power consumption for electrical sampling at the required sampling rate of 224 GSa/s for I and Q is estimated to be in excess of 160 W. This power value is calculated by interpolation from state-of-the-art analog-to-digital converters (ADC). A state-of-the-art ADC at 28 GSa/s consumes at least 10 W of electrical power. If higher sampling rates are implemented using parallelization, the power consumption increases linearly. If no guard interval is used, a sampling rate of 224 GSa/s on two ADCs is required leading to the estimated total power of 160 W. Introducing a guard interval at the same overall bitrate would increase the power consumption of the electronic implementation as digitizing of the guard interval is also needed. In comparison to this, the power consumption of the optical implementation will not increase, as the number of required optical gates does not change with the introduction of a guard interval. It has to be pointed out, that the power consumption for the electronic sampling also includes the analog-to-digital conversion.

A disadvantage of this approach is the unfortunate scaling with size. The number of couplers is the complexity $C_{\mathrm{std}} = N - 1 + (N/2)\log_2 N$, and the optical phases in all $N\log_2(N)$ arms of the FFT structure must be stabilized with respect to each other, thereby limiting N to a small number for practical cases. This renders the optical approach according to Fig. 5.1(c) impractical for large N. Electronic signal processing that could be used instead is, however strongly limited due to its power consumption and its limited speed. However, it is possible to significantly simplify the circuit of Fig. 5.1(c) without affecting its operation.

It has to be mentioned that another possible scheme uses waveguide grating routers (WGR) to implement the DFT [110, 111]. This approach, however, suffers from the need to control the relative phases of all N paths of the structure simultaneously, requiring N-1 Phase shifters.

5.1.2.2 A new optical FFT scheme

Here we show that by re-ordering the delays and by re-labeling the outputs accordingly an equivalent but simpler implementation can be found, since a direct implementation of the circuit in Fig. 5.1(c) would be difficult to make due to its frequent waveguide crossings, and due to the large number of waveguide phases that need to be accurately controlled.

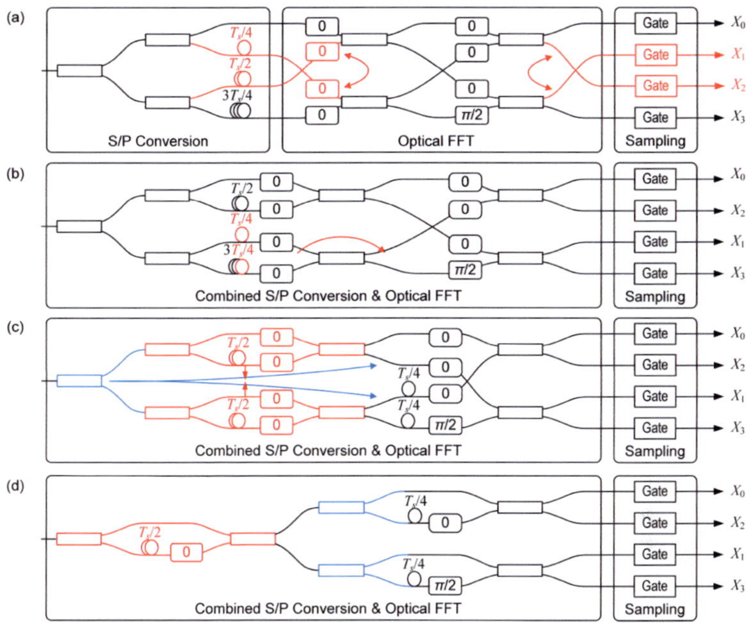

Fig. 5.2. Exemplary four-point optical FFT for symbol period T_s; (a) traditional implementation as in Fig. 5.1; (b) leading to a structure consisting of two DIs with the same differential delay; the additional $T_s/4$ delay is moved out of the second DI (c), which leads to two identical DIs that can be replaced by a single DI followed by signal splitters; (d) low-complexity scheme with combined S/P conversion and FFT.

The simplifying steps for an example with $N = 4$ are shown in Fig. 5.2. In a first step we relocate the sampling gates to the end of the circuit. This will not change the overall operation. Next we re-order the delays in the S/P conversion stage as indicated in Fig. 5.2(a) and re-label the outputs accordingly. This way the OFFT input stage consists of two parallel delay interferometers (DI) with the same free spectral range (FSR) but different absolute delays (cf. Fig. 5.2(b)). By moving the common delay of $T_s/4$ in both arms of the lower DI to its outputs, one obtains two identical DIs with the same input signal, see Fig. 5.2(c). This redundancy can be eliminated by replacing the two DIs with one DI and by splitting the output. The process is illustrated in Fig. 5.2(d). These simplification rules can be iterated to apply to FFTs of any size N. The new optical FFT processor consists only of $N-1$ cascaded DIs with a small complexity of only $C_{\mathrm{DI}} = 2(N-1)$ couplers, where $C_{\mathrm{DI}} \leq C_{\mathrm{std}} \; \forall N$. Also, in this implementation only the phase of $N-1$ DIs needs stabilization, and no inter-DI phase adjustment is required.

In the Appendix it is shown mathematically that the DFT of order $N = 2^p$ as described by Eq. (5.1) can always be replaced by an arrangement of p DI stages as indicated in Fig. 5.2(d). The DI delay in each stage and the location of the necessary phase shifters are derived as well. For illustration we show the structure for $N = 8$ in Fig. 5.3(a) as derived with the optical FFT approach according to [106] and in Fig. 5.3(b) the new structure presented in this contribution.

An inherent advantage of this approach is that a single frequency component of the sampled signal can be easily extracted without the implementation of the complete structure. In this case all DIs that are not part of the optical path to the corresponding output port can be removed, leaving only one DI per stage and thus a total of $\log_2 N$ DIs that require stabilization. By tuning the phases in each DI, any arbitrary FFT coefficient of the signal can be selected without changing the structure of the setup, as illustrated in Fig. 5.4. Such a reduction in complexity is not possible with the structures that have been shown previously (see Section 2.1). The ability to shape the spectral (Fourier) components of an optical signal with a structure similar to that in Fig. 5.4 makes it a member of the family of *Fourier filters* [112]. However, in order to obtain the block-wise DFT of an optical signal, a Fourier filter is not sufficient. Instead, the complete structure according to Fig. 5.3 is required, including the time-domain optical sampling which is an essential part of the DFT or short-time Fourier Transform (STFT), as we will show.

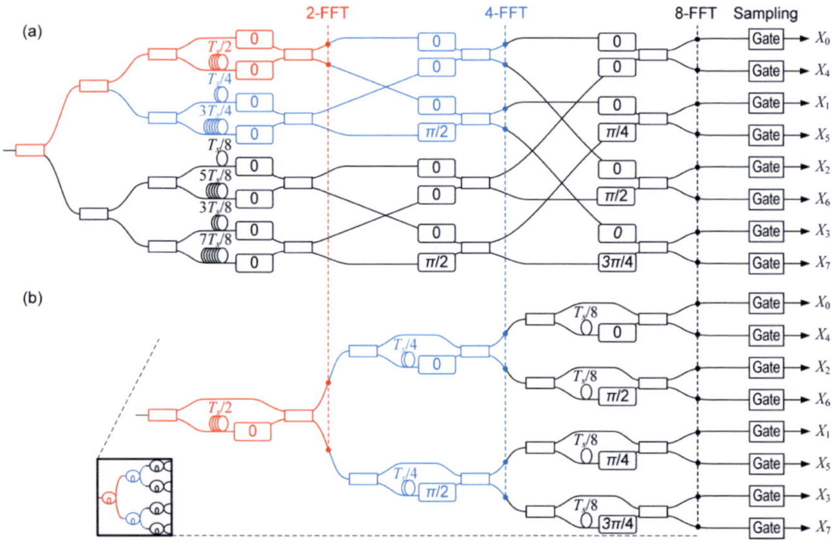

Fig. 5.3. (a) Direct FFT implementation versus (b) simplified all-optical FFT circuit for $N = 8$ showing the arrangement of delays and phase shifts as derived in Appendix A. The order of the outputs is different from that of the conventional FFT scheme. The sub-circuits for the FFT of order 2 and 4 are also marked, respectively. It should be mentioned, that a small misprint for one of the phase elements has been corrected from the published version [10].

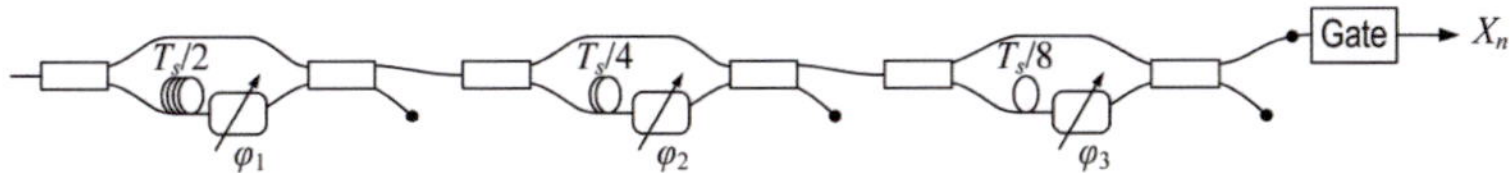

Fig. 5.4. Optical FFT circuit based on Fig. 5.3 for the extraction of Fourier component X_n. By tuning the phases φ_1, φ_2 and φ_3, any required Fourier component may be extracted without physically changing the setup.

The frequency response of the DI cascade can be easily visualized, cf. Fig. 5.5. As a matter of fact, the DFT of order N is able to discriminate N individual frequency components of the input signal, spaced $\Delta\omega = 2\pi/T_S$ apart, as theoretically shown in the Appendix. Thus, by cascading a sufficient number of DIs with correct delay and phase, any arbitrary frequency component can be isolated. Fig. 5.5 shows that switching between outputs X_0 and X_4 can be achieved by simply tuning $\varphi_3 = 0$ to $\varphi_3 = \pi$, thus shifting the green curve, equivalent to using the second (lower) output of the DI.

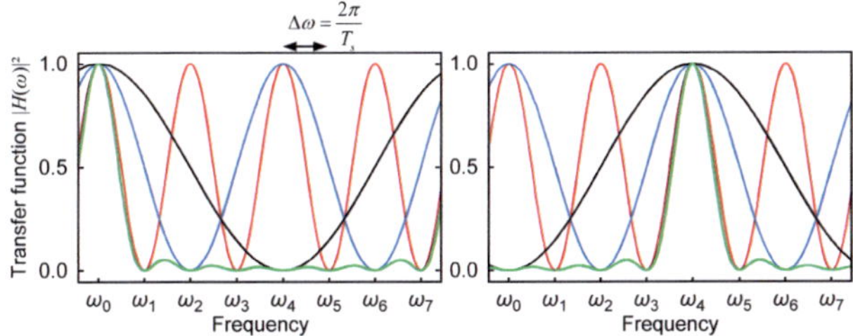

Fig. 5.5. Exemplary illustration of the intensity transfer functions of each stage (blue, red, and green) in the cascade of Fig. 5.3 and the total transfer function of the FFT circuit (black) for outputs X_0 (left side) and X_4 (right side).

The cascade of DIs has previously been proposed as a demultiplexer for FDM channels that do not overlap, functioning as a filter bank [110, 113-116]. Its adaptation to OFDM, which is based on the STFT/FFT property of the structure, however, is only possible in combination with optical sampling to delineate the OFDM symbol boundaries.

The traditional optical FFT according to Marhic [106] supplemented by optical sampling has also been implemented as OFDM demultiplexer by Takiguchi et al. [117] for $N = 4$ and recently even for $N = 8$ [118]. However, due to the complexity of the approach it does not scale well for larger N.

5.1.2.3 Mathematical Proof of the Scheme

Here we show that the DFT / FFT is equivalent to a cascade of DI. We show that the FFT has exactly the frequency response of the DIs.

The DFT in eq. (5.1) can be rewritten for continuous input and output signals $x(t)$ and $X_m(t)$

$$X_m(t) = \frac{1}{N} \cdot \sum_{n=0}^{N-1} \exp\left(-j2\pi n \frac{m}{N}\right) \cdot \delta\left(t - n\frac{T_s}{N}\right) * x(t) \qquad (5.3)$$

where $\delta(\cdot)$ is the Dirac delta function used to sample the input signal $x(t)$ at N equidistant points within the interval T_s by means of the convolution operation ($*$). Due to the numerous convolutions occurring when calculating the impulse response of cascaded elements, it is more appropriate, and equally valid, to do the comparison in the frequency domain. The transfer function $H_m(\omega)$ for the DFT can be obtained by a Fourier transform of (5.3),

$$\hat{X}_m(\omega) = \underbrace{\frac{1}{N} \cdot \sum_{n=0}^{N-1} \exp\left(-j2\pi\frac{nm}{N}\right)\exp\left(-j\omega\frac{nT_s}{N}\right)}_{H_m(\omega)} \cdot \hat{x}(\omega) \tag{5.4}$$

In which the caret ($^$) denotes the Fourier transform. It can be split into two sums corresponding to even and odd n,

$$H_m(\omega) = \frac{1}{N} \cdot \sum_{n=0}^{\frac{N}{2}-1}\left\{ \exp\left(-j2\pi[2n]\frac{m}{N}\right)\exp\left(-j\omega[2n]\frac{T_s}{N}\right) \right.$$
$$\left. + \exp\left(-j2\pi[2n+1]\frac{m}{N}\right)\exp\left(-j\omega[2n+1]\frac{T_s}{N}\right)\right\} \tag{5.5}$$

which can be simplified to

$$H_m(\omega) = \underbrace{\frac{2}{N} \cdot \sum_{n=0}^{\frac{N}{2}-1} \exp\left(-j\frac{2n}{N}[2\pi m + \omega T_s]\right)}_{\text{FFT of order } N/2} \cdot \underbrace{\frac{1}{2}\left[1+\exp\left(-j\left[\omega\frac{T_s}{N}+\frac{2\pi m}{N}\right]\right)\right]}_{H_{pm,1}(\omega)} \tag{5.6}$$

Here, $H_{pm,1}$ is the n-independent transfer function for the upper input of a delay interferometer with delay

$$T_{sa} = \frac{T_s}{N} = \frac{T_s}{2^p} \tag{5.7}$$

and additional phase shift

$$\varphi_{pm} = 2\pi\frac{m}{N} - \pi \tag{5.8}$$

obtained from the cascade of directional couplers and a delay line in the lower arm,

$$\mathbf{H}_{pm}(\omega) = \underbrace{\frac{1}{\sqrt{2}}\begin{pmatrix} 1 & j \\ j & 1 \end{pmatrix}}_{\text{output coupler}} \cdot \begin{pmatrix} 1 & 0 \\ 0 & \exp[-j(\omega T_{sa} + \varphi_m)] \end{pmatrix} \cdot \underbrace{\frac{1}{\sqrt{2}}\begin{pmatrix} 1 & j \\ j & 1 \end{pmatrix}}_{\text{input coupler}} \underbrace{\begin{pmatrix} 1 & 0 \\ 0 & 0 \end{pmatrix}}_{\substack{\text{upper input} \\ \text{isolation}}} \tag{5.9}$$

The remainder of expression (5.6) is the transfer function of the DFT of order $N/2$. Hence, a DFT of order N, with $N = 2^p$, can be implemented optically by cascading a DFT of order $N/2$ and a delay interferometer with delay T_s / N and output-specific phase shift φ_{pm}. It can be easily verified that the DFT transfer function for $N = 2$ is equal to both outputs of a single DI.

The DI phase φ_{pm} for an upper arm output X_m is the same as that for the lower arm output $jX_{m+N/2}$. The term describing the $N/2$-order FFT in (5.6) is also the same for X_m and $jX_{m+N/2}$ due to its periodicity. Thus both outputs of a single DI can be used to obtain different coefficients of the DFT, resulting directly in the optical FFT scheme of Fig. 5.3.

5.1.2.4 A further simplification of the optical FFT

Fig. 5.5 has shown that the DFT acts as a periodic filter in the frequency domain with a FSR of $N\Delta\omega$, and each DI of the cascade is also a periodic filter with FSR $N\Delta\omega/2^p$ where p is the index of the FFT stage and N is the order of the FFT. In order to further simplify the optical FFT circuit, one might be tempted to replace one or more stages of the DI cascade by standard (non-DI) optical filters. Since the stages with a higher subscript (those being traversed last) have the largest FSR, it would be sensible to replace these first, because the requirements on these filters are most relaxed. In Fig. 5.6 we have reduced the number of DI stages of an $N = 8$ FFT from top to bottom and replaced them by a single Gaussian filter (for illustration purposes).

Fig. 5.6(a) shows the original optical FFT with $N = 8$ together with the impulse response of output X_4. With the derivation given in Eq. (5.3) of the Appendix the impulse response at output X_4 is

$$h_4(t) = \frac{1}{8}\left[\sum_{n=0}^{3}\delta\left(t - \frac{2n+1}{8}T_s\right) - \sum_{k=0}^{3}\delta\left(t - \frac{2n}{8}T_s\right)\right] \tag{5.10}$$

and the corresponding frequency transfer function is

$$H_4(\omega) = \frac{1}{8}\left[1 + \exp\left(-j\left[\omega\frac{T_s}{2}\right]\right)\right]\left[1 + \exp\left(-j\left[\omega\frac{T_s}{4}\right]\right)\right]\left[1 + \exp\left(-j\left[\omega\frac{T_s}{8} + \pi\right]\right)\right] \tag{5.11}$$

If the third OFFT stage is replaced by a Gaussian filter (bank) as in Fig. 5.6(b), the FFT order is reduced to $N = 4$, which reduces the OFFT to two DI stages only. The impulse response then is described by the convolution of the response of the first two stages, which consists of 4 impulses according to Eq. (5.3) given in the Appendix, with the impulse response of a 1st-order Gaussian filter centered at ω_F,

$$h_{\text{Gauss}}(t) = \frac{1}{\sqrt{2\pi}\cdot\delta}\exp\left(-\frac{t^2}{2\delta^2}\right)\exp\left(-j\omega_F t\right) \quad \text{with} \quad \delta = \frac{\sqrt{\ln 2}}{\omega_B T} \tag{5.12}$$

where ω_B is the 3 dB bandwidth of the filter. As can be seen in the frequency response in Fig. 5.6(b), this replacement causes crosstalk from frequency component X_0 (red arrow). Furthermore, due to the significantly longer impulse response of the Gaussian filter, the total filter impulse response is increased (marked red) and causes inter-symbol interference (ISI). If more DI stages are replaced by optical filters, the filter passband in the frequency domain must become narrower. By the Fourier uncertainty principle [71] this results in an even longer impulse response and thus larger ISI, as shown in Fig. 5.6(c) and (d).

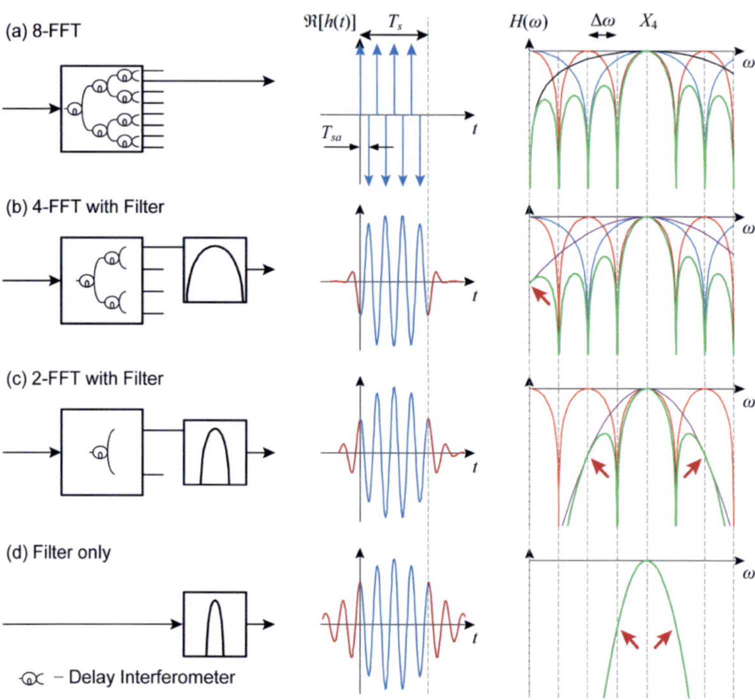

Fig. 5.6. Inter-symbol interference and frequency crosstalk occurring when replacing (parts of) the DI filters by 1st-order Gaussian filters with appropriate passbands in order to extract frequency component X_4. The left column shows the setup schematic, the middle column shows the real part of the impulse response and the right column shows the logarithmic intensity transfer functions of the involved DI stages (blue, red, and green), the optical filter (purple) and their cascade (black). In case of the approximations, the transfer function is not nulled for all outputs except X_4, leading to frequency crosstalk (red arrows). Also with decreasing filter bandwidth, the impulse response exceeds the DFT summation interval T_s (marked red), leading to crosstalk/interference from neighboring time slots.

5.1.2.5 Implementation

The all-optical (I)FFT structure presented in the previous section can be implemented straightforwardly using free-space optics or integrated optical circuits. Previously, implementations of DI cascades for up to $N = 16$ have been shown using the silica-on-silicon approach [113-116]. Integration in material systems of more recent interest such as silicon-on-insulator or InP is expected to yield much more compact structures, but has to our knowledge not yet been published. Benefits of optical integration are the small footprint and ease of stabilization.

5.1.3 Application to orthogonal frequency–division multiplexing (OFDM)

OFDM is a multi-carrier signaling technique that has emerged as a promising technology for ultra-high bit rate transmission. The reason lies in its potentially high spectral efficiency, which can be significantly higher than wavelength-division multiplexing (WDM), and its tolerance to transmission impairments like dispersion [119, 120]. A detailed overview of the state-of-the-art concerning OFDM transmission is available in [37].

In OFDM, the tributaries or subcarriers are spaced so tightly that their spectra overlap, whereas in WDM they are separated by guard bands which enable channel extraction by means of conventional optical filters, as shown in Fig. 5.7. The whole of the subcarriers in an OFDM channel form a signal in time – the OFDM signal, which can no longer be demodulated by a simple filter due to the spectral overlap of its tributaries. Its shape in time is an analog signal as was illustrated in Fig. 5.1. Yet if the subcarrier frequency spacing $\Delta\omega$ is related to the duration T_s by

$$\Delta\omega = \frac{2\pi}{T_s} \tag{5.13}$$

then two subcarriers k and l are orthogonal with respect to integration over an interval T,

$$\frac{1}{N}\sum_{n=0}^{N-1}\exp\left[jk\Delta\omega\frac{nT_s}{N}\right]\exp\left[jl\Delta\omega\frac{nT_s}{N}\right] = \delta_{kl}\,,\; \delta_{kl} = \begin{cases} 1 & \text{if } k=l \\ 0 & \text{else} \end{cases}. \tag{5.14}$$

As a consequence, an appropriate receiver will be able to distinguish them. Such receivers exist. They almost exclusively perform the FFT/STFT on the time-sampled signal in the electronic domain [121]. Such electronic real-time implementations are currently restricted to OFDM symbol rates of a few MBd due to speed limitations of the digital signal processor [122, 123]. These symbol rates correspond to total bitrates of several Gbit/s as a large number of subcarriers and higher order modulation formats are used. Higher bit rate OFDM signals usually have to be processed offline, which may be practicable for laboratory experiments but not for data transmission [83].

In this section we propose our low-complexity scheme based on the optical FFT, in which the demultiplexing of the OFDM subchannel is performed in the optical domain, and only the subchannel signal processing is done in the electronic domain. This way it is only the subchannel symbol rate that is limited by the capabilities of the electronic receiver.

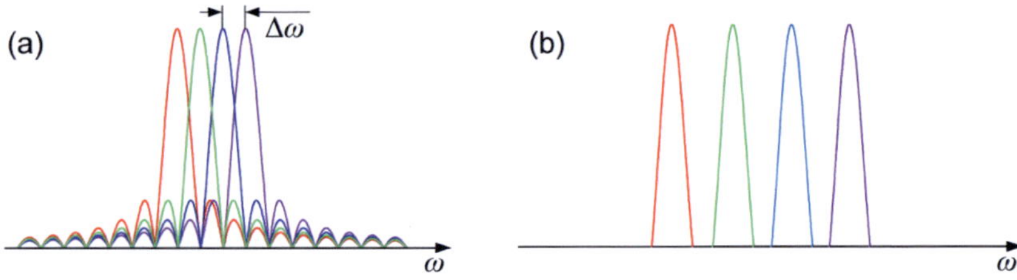

Fig. 5.7. Optical spectra of (a) an OFDM signal with 4 subchannels and (b) a DWDM signal with 4 channels. Due to the overlap, the OFDM subchannels cannot be extracted by simple optical filtering and the whole of the spectrum has to be processed simultaneously.

Two possible implementations exploiting the OFFT are depicted in Fig. 5.8. In Fig. 5.8(a), the optical inverse FFT (IFFT) is used to generate the OFDM signal and the optical FFT is used for demultiplexing. Here, the pulse train at rate T^1 from the mode-locked laser (MLL) is split and modulated independently with an arbitrary subcarrier modulation format for each subcarrier. The modulated pulse trains are subsequently fed into the optical IFFT circuit, which transforms them into an OFDM signal that consists of individual pulses of a length of $\sim T_{sa}$, not unlike an OTDM signal [33, 124]. Each of the output pulses of the optical FFT is a superposition of copies of the different input pulses with different phase coefficients. The spectrum of the generated signal is significantly wider than expected for an OFDM signal. To obtain the same waveform and spectrum as in the second approach, it is necessary to limit the bandwidth of the signal using bandpass filtering of the output signal or pulse shaping of the MLL. The main difference to an OTDM signal, which consists of a series of pulses of (ideally) equal amplitude, is that each pulse is a superposition of N subchannel samples with a corresponding variation in pulse amplitude, and the combination of N such pulses forms a single OFDM symbol. This transmitter is described in detail in Section 5.1.3.1. In Fig. 5.8(b), a frequency comb, here provided by a MLL is split into its (non-overlapping) Fourier components by a waveguide grating router (WGR). The Fourier components of the input signals are directly modulated at a symbol rate chosen such that condition (5.14) is fulfilled. The tributaries, or subchannels, are then recombined to obtain the OFDM signal. This transmitter is described in detail in Section 5.1.3.2.

In both cases, the receiver part consists of the optical FFT to demultiplex the OFDM signal into its tributaries and a subcarrier receiver "subchannel Rx". The exact details of the subcarrier Rx in Fig. 5.8 varies with the modulation format used within the subchannels. It could be a direct detection receiver, a balanced DI receiver such as needed for DPSK signals, a DQPSK receiver, or a coherent receiver for QPSK or any other QAM signal.

Comparing OFDM and OTDM at similar bitrates, one can observe some similarities, but also significant differences. Probably the most important ones can be found in the resilience with respect to chromatic dispersion. In OFDM systems, the dispersion tolerance can be tuned by a proper choice of subchannel bandwidth, and by the insertion of a cyclic prefix. As we have shown in Section 5.1.2.4, the extraction of an OFDM subband before demultiplexing can further increase the dispersion tolerance. On the other hand, OTDM requires that narrow symbols remain narrow, and thus requires higher-order dispersion compensation in order to be properly demultiplexed [33, 124]. Also, it is not possible to access a fraction of the OTDM signal using optical filtering, as it can be done for the in the subband access in OFDM signals. Lastly, OFDM requires the various lines of the comb source spectrum to be locked relative to each other only in frequency, whereas for OTDM (and the IOFFT transmitter of Section 5.1.3.1) the spectral lines need to obey strict phase relations in order to obtain sufficiently short pulses.

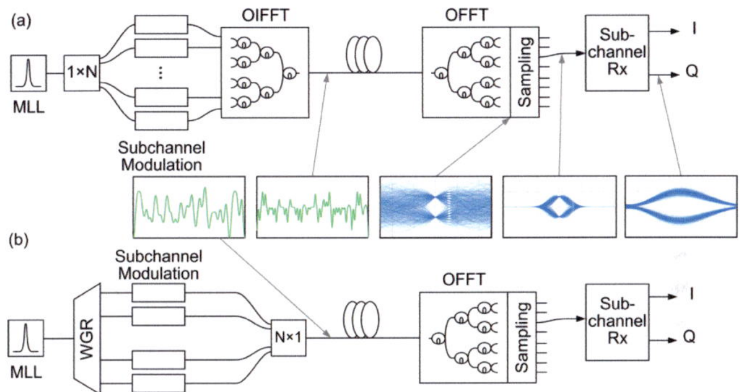

Fig. 5.8. Two examples for the implementation of an all-optical OFDM transmitter-receiver pair. (a) The output of a pulse source (e.g. a mode-locked laser) is split onto N copies and each copy of the pulse train is encoded individually with an arbitrary subcarrier modulation format before being combined in the optical IFFT circuit. (b) The output of a pulse or frequency comb source is split into its spectral components, each of which satisfies eq. (5.14). Those spectral components are separately encoded with an arbitrary subcarrier modulation format and combined to form the OFDM signal. At the receiver, the subchannels are separated using the optical FFT circuit. The receiver is identical to the one above. Green insets show exemplary waveforms before WDM filtering at the transmitter with OTDM-like pulses after the optical IFFT. Blue insets show exemplary eye diagrams at various locations within the receiver.

5.1.3.1 OFDM using the optical IFFT

The optical FFT can be used for both, the implementation of the transmitter and of the receiver, Fig. 5.8(b). At the transmitter, the IFFT creates the OFDM symbols with the configuration of Fig. 5.3, whereas the signals would traverse the structure in the inverse direction, coming from the right. A short pulse launched into any of the IFFT circuit inputs will result in a series of pulses with equal shapes but correspondingly lower pulse energy to appear at the output of the circuit. In order for these pulses not to interfere with one another or with pulses from neighboring OFDM symbols, the duration of these input pulses must be sufficiently short (on the order of T_s/N). The input pulses can then be encoded with an arbitrary modulation format (e.g. QPSK or QAM) prior to injection into the optical IFFT circuit to obtain a sequence of OFDM symbols. At the receiver, the optical FFT circuit demultiplexes the OFDM symbols continuously. Correct outputs are obtained only when the FFT window is synchronized with the OFDM symbols. Otherwise intersymbol interference (ISI) will occur. This reduces the usable width of the received signal (the open "eye") at the receiver by a factor of approximately N. To extract the usable ISI-free time slot of the received signal, an optical gate must be part of the FFT circuit, such that bandwidth-limited receiver electronics may be used [110].

In OFDM transmission, accumulated group-velocity dispersion (GVD) will introduce crosstalk by shifting the OFDM symbol boundaries in each subchannel, which causes blurring [120, 121]. However, since subchannel rates and spacing can be high in our scheme, the sensitivity to GVD is non-negligible. The addition of a cyclic prefix could alleviate the problem, but is difficult to realize within the optical IFFT circuit, as a part of the optical signal would need to be duplicated and delayed increasing the complexity of the scheme.

5.1.3.2 OFDM using a frequency comb

At the transmitter, the subchannel rate limitations imposed by electronics may be overcome by using a DWDM-like approach, where the possibility to optically generate precisely tuned spectral components in frequency space is exploited to directly generate OFDM subcarriers at the correct frequency separation $\Delta\omega$ as required by the OFDM condition (5.13). A bank of frequency offset-locked laser diodes or an optical comb generator provide subchannel carriers which can be modulated individually [125]. Also the concept of a recirculating frequency shifter (RFS) has been successfully applied to generation of frequency offset-locked subcarriers [48]. A similar approach is used by coherent WDM systems, which forego the OFFT at the receiver in favor of standard optical filters, but require a phase-synchronized transmitter laser bank [45].

In the approach shown in Fig. 5.8(a), the frequency comb of a pulse source is spectrally separated into subchannel carriers. Each subcarrier, frequency-locked but not phase-synchronized, is then individually modulated and combined to form OFDM symbols. Hence, a corresponding optical FFT receiver can be used to decode the subchannels. The major difference to the IFFT transmitter is that the output corresponding to any one input is not a series of pulses with discrete phases, but a continuous signal with a corresponding optical frequency. This transmitter can be considered as the continuous Fourier transform equivalent of the discrete transform performed by the IFFT.

In such a transmitter, bandwidth limitations of the modulator will cause subchannel crosstalk because the orthogonality condition (5.14) cannot be fulfilled in the presence of residual amplitude modulation of the subchannels near the symbol boundaries. The orthogonality condition actually forbids residual modulation of the subcarriers within the FFT window – the phase and amplitude of the complex signal to be encoded must be maintained throughout the symbol time T_s. This would require the transition from one OFDM symbol to the next to be instantaneous, requiring modulators of infinite bandwidth. Any transition region between adjacent symbols therefore would lead to a violation of the orthogonality condition. If the subchannels are not orthogonal, crosstalk will be generated when performing the FFT in the receiver. As a result, the portion of the symbol duration usable for detection at the receiver – and thus the width of the gating window is shortened. This is illustrated in Fig. 5.9(a). Similar to traditional OFDM [37], a remedy is the insertion of a guard interval (corresponding to the cyclic prefix) between symbols so that the *FFT window duration at the receiver will not be changed.* By making the guard interval just long enough to contain the intersymbol transition, the remaining part of the OFDM symbol fulfils the orthogonality condition and receiver crosstalk will only be generated within the guard interval. The width of the open "eye" at the receiver increases, albeit at the cost of a slightly reduced symbol rate. This is illustrated in Fig. 5.9(b) and (c).

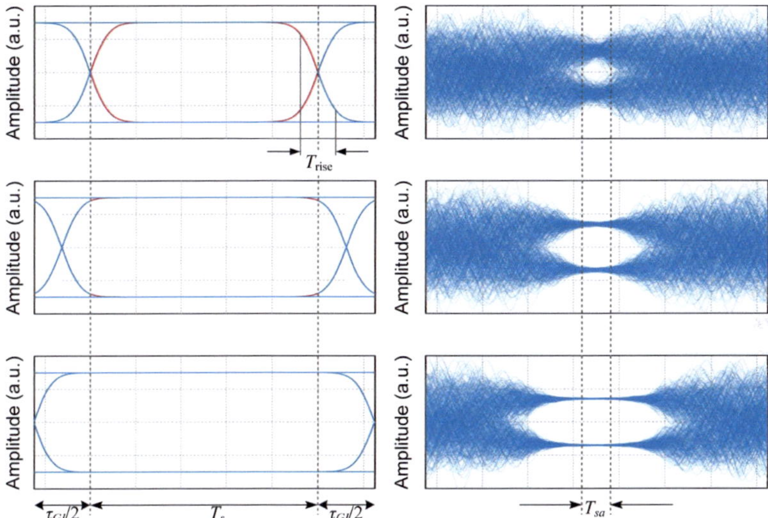

Fig. 5.9. Function of an OFDM guard interval using the setup of Fig. 5.8(b) and an 8-FFT. On the left-hand side, the exemplary eye diagram of a single modulated subchannel is shown, as is, on the right side, the received signal before optical gating. If the symbol duration equals the integration interval T_s, the signal transitions, described by the 10-90% rise/fall time T_{rise} (red), cause inter- and intra-subchannel crosstalk and the received eye is almost fully closed within the observation window of length $T_{sa} = T_s/N$. With increasing length of the guard interval τ_{GI}, interference vanishes during T_s. Further increasing the guard interval increases the duration in which the orthogonality condition is fulfilled and thus increases the duration of the open "eye."

An additional increase of the guard interval increases resilience towards accumulated GVD. For an increase of the guard interval length τ_{GI} (beyond that necessary to achieve the required degree of orthogonality) one can keep up with a maximum accumulated GVD B_2,

$$B_2 = \int_{z=0}^{L} \beta_2 \, dz = \frac{\tau_{GI}}{B_{OFDM}} \tag{5.15}$$

where L is the system length, β_2 is the local GVD coefficient and B_{OFDM} is the bandwidth of the OFDM (super-)channel. By using optical filters to extract a slice of the OFDM channel before performing an FFT of correspondingly lower order, as shown in Fig. 5.6, the bandwidth $\Delta\omega$ over which GVD may introduce crosstalk is reduced. Thus the resilience towards accumulated GVD increases (see Fig. 5.10).

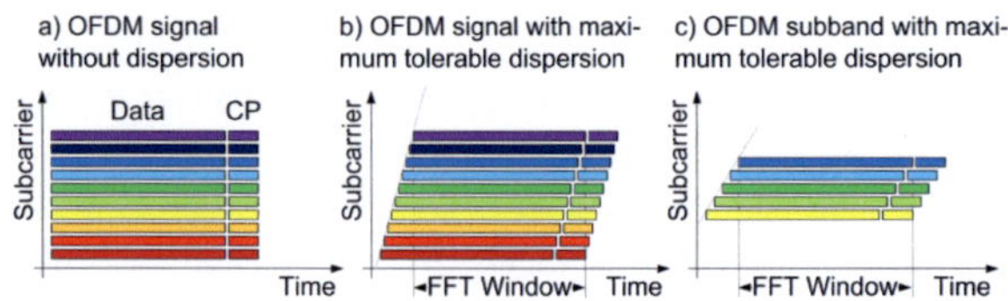

Fig. 5.10: Mitigating dispersion by means of a cyclic prefix (CP). (a) An OFDM signal with data and cyclic prefix. (b) Frequency dependent delay of subcarriers after transmission due to dispersion. The amount of dispersion that can be tolerated is limited by the length of the cyclic prefix (CP) as all subcarrier symbols must stay within the FFT window. (c) By means of optical filtering, an OFDM subband has been extracted, as discussed in Fig. 5.6. This way, only the extracted symbols must stay within the FFT window, and thus a larger amount of dispersion can be tolerated.

However, as discussed earlier, it comes at a price. Conventional optical bandpass filters introduce inter-subchannel crosstalk and inter-symbol interference and thus reduce the quality of the received signal. Nevertheless, such a scheme has been successfully used in [45] to extract a single subchannel of an optically generated OFDM signal. The combination of a cascade of DIs and simple optical filters to perform OFDM demultiplexing has also been exploited in the electrical domain. In [48], cascaded delay-and-add elements were used instead of an FFT in order to extract single subchannels out of an optically filtered section of the OFDM channel spectrum. This method reduces cost in terms of time and complexity, but is still limited by the speed of electronics. Another approach to reduce the required electronics speed is multi-band OFDM [40, 61], which sacrifices some spectral efficiency for the ability to extract OFDM subbands by tuning the LO laser within the receiver.

Exemplary, Fig. 5.11 shows the dependence of the back-to-back received signal quality, determined in terms of Q_F,

$$Q_F = \frac{\overline{I}_0 - \overline{I}_1}{\sigma_0 + \sigma_1} \tag{5.16}$$

averaged over the in-phase and quadrature components of the central subchannel, versus the bandwidth of a 4th-order Gaussian filter for various simplifying implementations of the optical FFT and for a guard interval of 25 % of the OFDM Symbol duration T_s. In eq. (5.16), the quantities $\overline{I}$ denote the mean signal values and the symbols σ mean the standard deviations of the received symbol levels (in this case the in-phase and quadrature values of QPSK subchannel signals). The transmitter setup shown in Fig. 5.8(c) was used with a modulator rise time $T_{\text{rise}} = T_s/8$ and an optical 200 GHz bandpass filter following the modulator. Clearly, the signal quality deteriorates with the number of DI stages that have been replaced by conventional filters. However, the improvement in residual GVD tolerance may well offset the penalty in systems with a large number of subchannels.

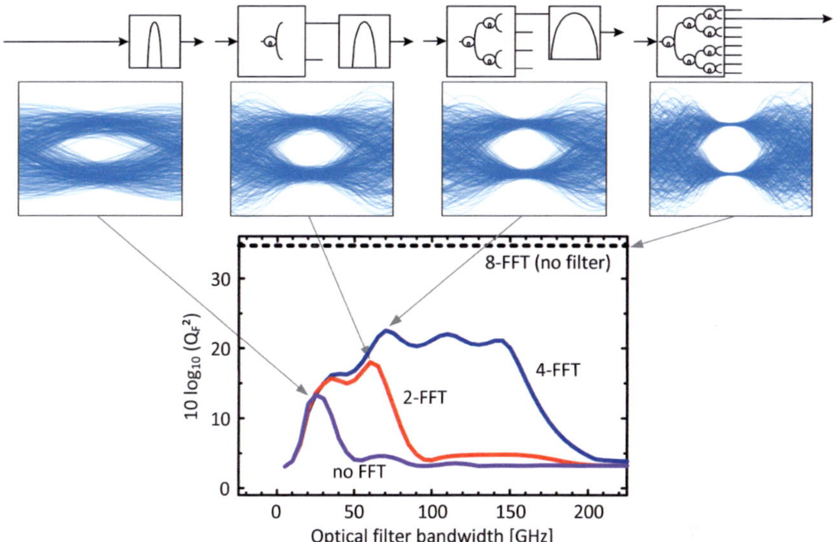

Fig. 5.11. Achievable quality of the received signal for various FFT filter schemes performed on an 8-channel OFDM signal with 20 GBd on a 25 GHz subcarrier spacing. The solid red plot shows the signal quality if the FFT is performed with a Gaussian filter as a function of the Gaussian filter bandwidth. The blue and green curves show the signal qualities if one and two DI cascades are used. The black curve shows how a perfect FFT can be performed if the FFT is performed with DIs only.

5.1.4 Summary

In [10] a practical scheme for optical FFT processing has been introduced. The implemented OFDM receiver shows no penalty compared to single channel back-to-back performance. As the scheme is not subject to any electronic speed limitations it will allow Tbit/s FFT processing. Also, since the scheme relies on passive optical filters it performs processing with virtually no power consumption and this way may help to overcome the ever increasing energy demand that normally comes with higher speed.

5.2 Proof–of–Concept Demonstration

To verify the feasibility of the all-optical OFDM generation, including a guard interval, and OFDM demultiplexing by means of the OFFT, a back-to-back experiment was performed, which will be briefly described in this section [9].

5.2.1 Experimental setup

The OFDM receiver and transmitter are shown in Fig. 5.12. The transmitter is based on the principles presented in Section 5.1.3.2. The frequency-locked subcarriers are generated by a 50 GHz comb generator providing 9 sufficiently strong OFDM subcarriers. The optical comb source is based on two cascaded dual drive Mach-Zehnder modulators that are driven by an electrical clock signal as demonstrated in [45]. These are separated into odd and even channels by a disinterleaver. The odd channels are encoded with PRBS 2^7-1 DPSK data at 28 GBd while the even channels are encoded with 28 GBd DQPSK data, using decorrelated PRBS 2^7-1 sequences for the in-phase and quadrature components. After combination within an optical coupler, the channel spectra overlap significantly and thus can no longer be demultiplexed by standard optical filters. This was verified with the help of simulations and experiments. We end up with a 392 Gbit/s OFDM signal.

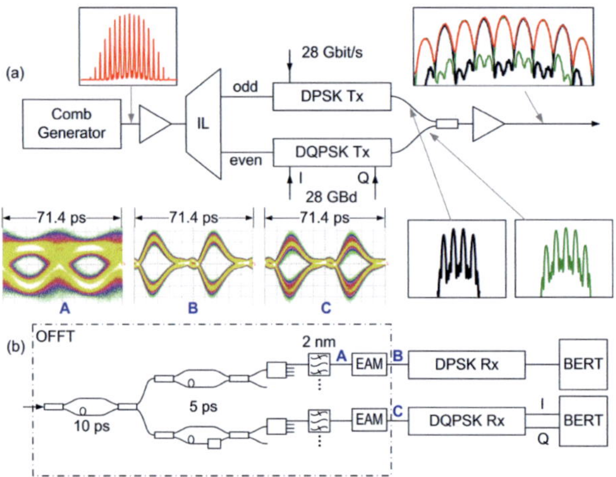

Fig. 5.12: Setup of OFDM transmission system with (a) transmitter and (b) receiver. Two cascaded Mach-Zehnder modulators generate an optical frequency comb, which is split by a disinterleaver into 4 odd and 5 even channels. Spectrally adjacent subcarriers are modulated alternately using DBPSK or DQPSK modulation. All subcarriers are combined in a coupler and transmitted. The received OFDM signal is processed using the low-complexity OFFT circuit of Section 5.1.2.4 with 2 DIs and one standard optical filter. The resulting signals are sampled by electro-absorption modulators (EAM) and detected using DBPSK and DQPSK receivers. Bit error rates were measured with a BERT. The receiver comprises the all-optical FFT scheme followed by a preamplified receiver with differential direct detection. The optical FFT circuit consists of a cascade of two DIs, followed by passive splitters and a bank of bandpass filters. We thus adopt the low-complexity FFT circuit of Section 5.1.2.4, partially compensating for the associated performance loss by the increased guard interval. The final elements of the OFFT are the EAM sampling gates. After OFDM demultiplexing, the signals are detected using standard DBPSK and DQPSK receivers. Either eye diagrams A, B, C or bit error probabilities BER (Fig. 5.13) were measured with a 50 GHz photodiode and a 70 GHz electrical sampling oscilloscope or a bit error tester (BERT), respectively. Spectra are plotted with 20 dB/div (vertically) and 2 nm/div (horizontally) in a resolution bandwidth of 0.01 nm, center of plotted spectra located at 1550 nm.

The guard interval length was set to 15.7 ps due to the significant rise and fall times of the optical modulator used, and to increase the sampling window size at the receiver. Had we been able to use components with a larger bandwidth and therefore shorter rise and falltimes, the guard interval could have been reduced significantly without affecting the operation of the optical FFT at the receiver, and thus bring the OFDM symbol rate closer to the subchannel separation frequency $\Delta\omega$.

The receiver comprises the new all-optical FFT scheme followed by a preamplified receiver with differential direct-detection. The FFT processor comprises a cascade of two DI, followed by passive splitters and bandpass filters (explained in the previous section), and the EAM sampling gates. As gate, an electro absorption modulator (EAM – CIP 40G-PS-EAM-1550) is chosen.

Intercarrier and intersymbol interference (see eye diagram **A**) are suppressed by the optical gate, which also sets the FFT window. Bit error probabilities (BER) for each subcarrier have been measured on the output signals (**B, C**).

To evaluate the OFFT performance, we compared subcarrier signals after multiplexing/demultiplexing with back-to-back (B2B) signals delivered by the DBPSK and DQPSK transmitters, respectively, in terms of bit error ratio (BER) versus receiver input power.

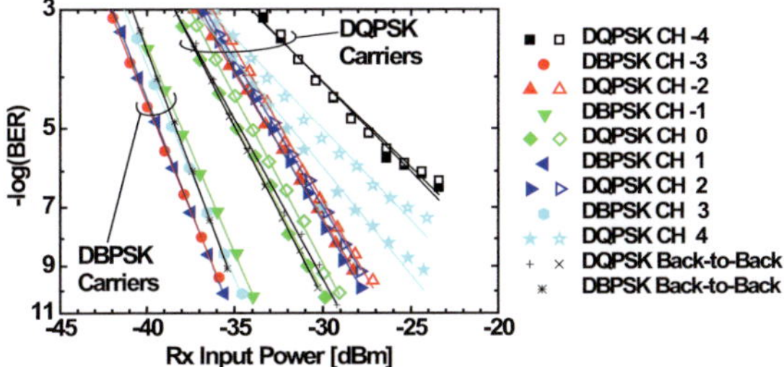

Fig. 5.13: BER performance of different subcarriers. No penalty is observed compared to back-to-back performance for DBPSK carriers (−3, −1, 1, 3) and no significant penalty for the central DQPSK carriers (−2, 0, 2). A 5 dB penalty or error floor occurs for the two outer DQPSK subcarriers (−4, 4), which are launched with 11 dB less power in the optical comb.

To measure a comparable B2B signal performance, the outputs of the transmitters were gated using the same EAM as it was used in the OFDM receiver. The results depicted in Fig. 5.13 show no penalty compared to the B2B performance for DBPSK and only a small penalty for the DQPSK channels, owing in part to the large guard interval. The outer channels (labelled "−4" and "4") perform comparatively worse, because these subcarriers are generated with 11 dB less optical power compared to the center channel. The capacity limit for an OFDM channel using this approach is therefore mainly determined by the performance of the frequency comb generator – the number of subcarriers which are generated and the signal-to-noise ratio with which this is done.

5.3 Conclusion

This is the first experimental demonstration of an all-optical FFT based on cascaded DI. The capabilities are demonstrated through the implementation of an OFDM receiver that is showing no penalty compared to single-channel B2B performance. The scheme enables ultra-fast OFDM transmission with state-of-the-art electronics.

6 Single–Laser Terabit/s Experiments

In this chapter is a summary of record experiments with the highest data rates on a single laser. Section 0 has been published in Nature Photonics [8] and section 6.2 has been published in [13].

6.1 All–Optical OFDM record experiment

26 Tbit s⁻¹ line–rate super–channel transmission utilizing all–optical fast Fourier transform processing

D. Hillerkuss, R. Schmogrow, T. Schellinger, M. Jordan, M. Winter, G. Huber, T. Vallaitis, R. Bonk, P. Kleinow, F. Frey, M. Roeger, S. Koenig, A. Ludwig, A. Marculescu, J. Li, M. Hoh, M. Dreschmann, J. Meyer, S. Ben-Ezra, N. Narkiss, B. Nebendahl, F. Parmigiani, P. Petropoulos, B. Resan, A. Oehler, K. Weingarten, T. Ellermeyer, J. Lutz, M. Moeller, M. Huebner, J. Becker, C. Koos, W. Freude, and J. Leuthold

Nature Photonics **5**(6), 364-371 (2011). [8]

Optical transmission systems with Terabit/s single channel line rates no longer seem to be too far-fetched [126]. New services such as cloud computing, 3D high-definition TV and virtual reality applications require unprecedented optical channel bandwidths. These high-capacity optical channels, however, are fed from lower-bitrate signals. The question then is if the lower-bitrate tributary information can viably, energy-efficiently and effortlessly be encoded to and extracted from Terabit/s (Tbit s⁻¹) data streams. Here we demonstrate an optical fast Fourier transform (FFT) scheme that provides the necessary computing power to encode lower-bitrate tributaries into a 10.8 Tbit s⁻¹ and 26.0 Tbit s⁻¹ line-rate orthogonal frequency division multiplexing (OFDM) data stream and to decode them from the fibre-transmitted OFDM stream. Experiments show the feasibility and the ease in handling Tbit s⁻¹ data with low energy consumption. To the best of our knowledge, this is the largest line-rate ever encoded onto a single light source.

Traditionally, time division multiplexing (TDM), in which multiple bit streams are multiplexed onto a single signal by assigning a recurrent timeslot to each of the tributaries, has increased the amount of information encoded on a single laser wavelength. Recently, data streams of 10.2 Tbit s⁻¹ have been encoded on a single laser by applying a TDM scheme[33]. Yet, TDM pulses are short and spectrally broad, see Fig. 6.1a. This makes TDM schemes challenging. The short pulses (e.g. [33], 300 fs) require a narrow receiver time window, and the large optical bandwidth (e.g.[33], 30 nm) makes precisely engineered dispersion compensation a necessity. Alternatively, with wavelength division multiplexing (WDM) large aggregated bitrates are transmitted over optical fibres [35]. In WDM, several wavelengths transport several data streams in parallel. Pulse durations are longer, and optical bandwidths

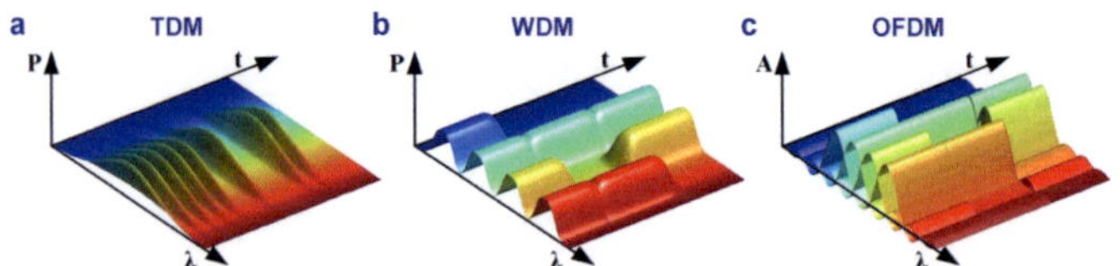

Fig. 6.1: Comparison of multiplexing schemes. **a** In time division multiplexing (TDM), data are transmitted in form of a serial stream of pulses. **b** For wavelength division multiplexing (WDM), data are distributed over several carrier wavelengths and transmitted in parallel. The different wavelength tributaries can be separated by optical bandpass filters. **c** In an orthogonal frequency division multiplexed (OFDM) modulation scheme, data are — like in WDM — transmitted on a number of subcarriers in parallel. However, as modulated OFDM subcarrier spectra overlap, simple bandpass filters cannot separate the tributaries. Therefore, signal processing at the receiver with e.g. a Fourier transform processing is required.

are moderate (Fig. 6.1b). Spectral guard bands are required to avoid crosstalk from one channel to the others. This reduces the spectral efficiency, and the dedicated receivers and transmitters with stabilized lasers for each transmission wavelength makes the systems expensive. OFDM is a more recent approach in optical communications [7, 36-40], while already well known from wireless data transmission. In contrast to WDM, modulated OFDM subcarriers overlap each other significantly, Fig. 6.1c. This way spectral efficiency is high, and the encoding of information on a large number of subcarriers makes OFDM tolerant towards dispersion [38-40]. A detailed summary and discussion of advantages and disadvantages can be found [37].

The classical method to generate an OFDM signal is shown in Fig. 6.2a. Inside a symbol time slot T_S we position for instance $N = 4$ independent data symbols. Each symbol X_m is interpreted as a complex Fourier coefficient in the frequency domain. Upon performing the inverse fast Fourier transform (IFFT) on these spectral data, one obtains complex time-domain samples x_n, which, after parallel-to-serial and digital-to-analogue conversion (DAC), make up a complex analogue OFDM signal $x(t)$. This analogue OFDM signal is subsequently modulated on an optical carrier and finally transmitted over an optical fibre. Unlike an ordinary signal, the spectrum of the OFDM signal comprises many subcarriers numbered by m which transmit the associated complex data X_m. Neighbouring subcarriers are separated by an angular frequency interval $\Delta\omega = 2\pi / T_S$. In contrast to WDM transmission, the spectrally overlapping subcarrier information can no longer be retrieved with standard bandpass filters. Instead, the process inverse to the one on the transmitter side, namely an analogue-to-digital conversion (ADC) followed by an FFT, yields the complex Fourier coefficients representing the transmitted data. Both the IFFT and FFT are typically performed in the electronic domain and are therefore limited in bitrate. Until now real-time electronic IFFT and FFT signal processing for OFDM signals up to 101.5 Gbit s^{-1} [31, 122, 123] has been demonstrated. This is far off from what would be desirable for the generation or reception of Tbit s^{-1} OFDM signals. An all-optical solution that could work beyond the speed limitations of electronics would therefore be of interest.

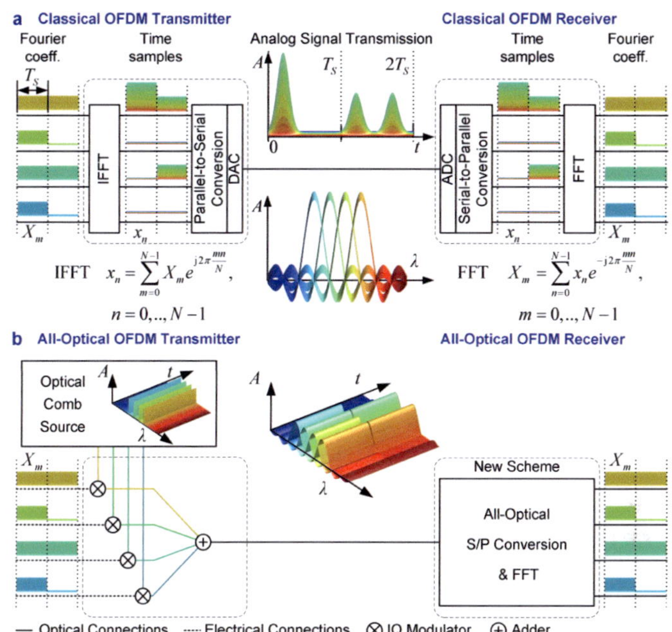

$$\mathrm{IFFT} \quad x_n = \sum_{m=0}^{N-1} X_m e^{j2\pi\frac{mn}{N}}, \qquad n = 0,..,N-1$$

$$\mathrm{FFT} \quad X_m = \sum_{n=0}^{N-1} x_n e^{-j2\pi\frac{mn}{N}}, \qquad m = 0,..,N-1$$

Fig. 6.2: OFDM processing. **a** Classical implementation: Here, N = 4 parallel data-streams are interpreted as complex Fourier coefficients in the frequency domain. The IFFT yields time-samples xn of a signal. Parallel-to-serial and digital-to-analogue conversion provides an analogue OFDM signal A(t) with spectrum A(λ). After transmission, the information is retrieved through the inverse process: An analogue-to-digital conversion (ADC) and a serial-to-parallel conversion followed by a FFT return the Fourier coefficients representing the data streams. **b** All-optical implementation: Data-streams are IQ-modulated ⊗ on equidistant, frequency-locked subcarriers. After summation, an OFDM signal is obtained as with the classical method. The all-optical OFDM-receiver is discussed elsewhere [10].

A straightforward implementation of the IFFT to generate an OFDM signal has recently been suggested, see the left-hand side of Fig. 6.2b. Individual laser sources at equidistant frequencies serve as subcarriers onto which IQ modulators encode the information to be transmitted. The modulated subcarriers are then combined with an optical coupler to form the OFDM signal [45, 127]. This will result in excess losses compared to the usual WDM technology.

Conversely, at the receiver side an optical FFT can be used to separate the subcarriers efficiently. Recently, we introduced an all-optical configuration [10], which performs both serial-to-parallel (S/P) conversion and the FFT in the optical domain using a cascade of delay interferometers (DI) with subsequent time gates (Fig. 6.3b). We obtained this configuration by combining and rearranging the elements of a direct optical FFT implementation previously suggested by Marhic [106]. A mathematical derivation showing that the new configuration represents indeed an FFT was given [10]. A generic derivation can be found [128]. The advantage of this configuration over other suggestions [106, 110, 111, 118, 129, 130] is that it requires only $N-1$ passive DIs and simple gating to perform an N-point FFT. It therefore scales well with the bitrate, and requires only a very small amount of energy for phase adjustments and gating. If only a single subcarrier needs detection the number of necessary DI stages further reduces to $\log_2 N$ as only one branch of the DI cascade is needed. It should be clarified, that for $N = 2$ our new implementation and the one by Marhic [106, 129] are identical. However, for $N > 2$ we obtain a significantly simpler setup. In addition, because the first few DI stages have the largest impact on the overall performance, only these stages are needed, and a number of DI at the end of the cascade can be replaced by bandpass filters, thereby simplifying the FFT processing even further [10]. This is because only closely neighbouring channels overlap to a significant amount. The optical time gates at the end of the FFT circuit (one per OFDM subcarrier) are operated in synchronism with the symbol rate. Its purpose is to select the window in time in which only samples from within one OFDM symbol contribute to the output of the optical FFT [10]. By adding a cyclic prefix to the symbol this time window widens accordingly. This way the time-width of the gate is not required to be as narrow as for an equivalent OTDM symbol duration T_S / N. The power consumption of the discrete optical gate actually is quite low (in our case 1.75 Watt per gate). This corresponds to a total power consumption of only 131.25 W for a setup built from discrete components that can process 10.8 Tbit/s in real-time. With electro-optic integrated circuits, the power could be greatly reduced.

In practice, the optical approach serves two purposes. First, it allows signal processing at a record aggregated bitrate with little energy consumption, and second, it rescales the bitrate BR of an incoming high-speed optical signal to its lower-speed tributary channels having bitrate BR/N. This is important for OFDM processing where analogue-digital converters at a line bitrate BR are prohibitive or unavailable, while being affordable and available at bitrates BR/N.

In the following, we use our FFT scheme to encode a 10.8 Tbit s^{-1} line rate OFDM signal onto a single laser beam and demonstrate real-time processing of the signal to recover its 75 tributaries [7]. To further prove the scalability and viability of the scheme, we also demonstrate the generation of a 26 Tbit s^{-1} signal with 325 modulated subcarriers. This OFDM signal is transmitted over a distance of 50 km and decoded by the all-optical FFT processor in the receiver.

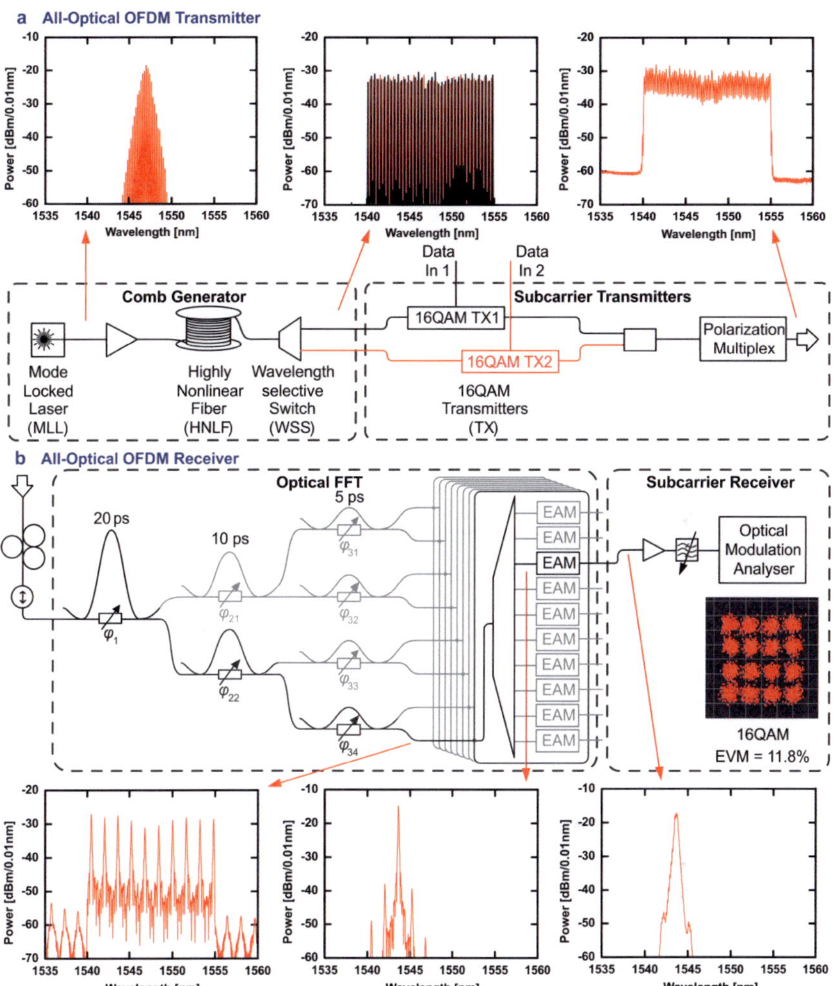

Fig. 6.3: All-optical OFDM transmitter and receiver. **a** All-optical OFDM transmitter: A comb generator (mode-locked laser (MLL) with highly nonlinear fibre) generates a frequency comb. A wavelength selective switch provides dis-interleaving into even and odd channels and equalization. All 75 equidistant subcarriers are individually modulated with 16QAM, recombined and polarization multiplexed to generate the OFDM signal. It should be mentioned that two insets were switched due to a misprint in [8]. This has been corrected in this figure **b** All-optical OFDM receiver: The optical FFT consists of a 3-stage delay interferometer (DI) cascade, one branch of which was actually implemented, two cascaded single-cavity 1 nm filters, and an EAM gate, after which the signal is received by an optical modulation analyser

6.1.1 Demonstration of the concept at a line rate of 10.8 Tbit s^{-1}

The OFDM transmitter and receiver setup are shown in Fig. 6.3. In the transmitter, an optical comb generator comprising a single mode-locked laser (MLL) directly generates OFDM sub-carriers with frequency separation $\Delta f = 25$ GHz. After a booster amplifier, a highly nonlinear fibre broadens the original MLL comb. A programmable optical filter equalizes the power of all optical subcarriers and separates them into even and odd subcarriers. Both sets of subcarriers are then individually modulated with independent 16QAM pseudo-random bit sequences (PRBS 2^{15}-1), and subsequently combined to form the OFDM signal [1, 3]. The OFDM orthogonality condition would require a symbol duration of $T_S = 1/\Delta f = 40$ ps [37], which would allow us to encode a symbol rate of up to 25 GBd. However, the bandwidth limitations of the modulator and the resulting rise and fall times inside T_S would cause the symbol not to be constant in amplitude and phase over the whole duration of the OFDM symbol. This would lead to inter-subcarrier crosstalk, which can be mitigated by inserting a cyclic prefix of $\tau_c = 15.6$ ps [10] to shift the rise and fall times of transmitter and receiver out of the symbol duration window. This corresponds to a cyclic prefix of CP $= \tau_c / (T_s + \tau_c) = 28\%$ [131], i. e., 28 % of the resulting reciprocal symbol rate. The cyclic prefix is inserted by modulating the subcarriers with a reduced symbol rate of 18 GBd (=25 GBd × 0.72; the symbol duration T_S, however, remains constant). To emulate a polarization-multiplexed signal, the transmitter output is split, one of the paths is delayed by 5.3 ns with respect to the other to decorrelate the data, and both paths are then united by a polarization beam combiner.

The receiver comprises the all-optical OFDM-FFT circuit [10] and a commercially available coherent receiver (Agilent optical modulation analyser - OMA) for down conversion, sampling and signal analysis. It utilizes a real-time oscilloscope with 13 GHz electrical bandwidth. One branch of an eight-point FFT is implemented using a cascade of DIs, followed by a cascade of two 1 nm bandpass filters to select one of the subcarriers extracted by the FFT. The final element of the OFFT is an electro-absorption modulator gate (EAM – CIP 40G-PS-EAM) that is operated at 18 GHz with 3 V$_{pp}$ modulation voltage, thereby generating a gating window of 15 ps FWHM (measured with a 70 GHz photodiode and 70 GHz electrical sampling heads). In this demonstration, the EAM was driven with the transmitter clock. For selecting the N samples for each symbol uniquely, the gating window cannot be larger than the symbol duration divided by the number of samples, here 40 ps / 8 = 5 ps, otherwise neighbouring symbols would interfere. The introduction of a cyclic prefix (15.6 ps in our case), however, increases the acceptable gate width to 20.6 ps. With regard to finite rise and fall times, we eventually choose a window width of 15 ps. A tunable polarization filter is inserted before the optical FFT circuit to perform polarization demultiplexing.

We then measured the error vector magnitude (EVM) [132] for both polarizations and all 75 subcarriers using the all-optical-FFT-based all-optical OFDM receiver with the help of the OMA (Fig. 6.4d). The signal quality measure EVM has been common for OFDM in mobile communications for a long time. The relation between EVM and BER can be found in the methods section. Each measurement comprises 2^{10} received symbols. A typical 16QAM constellation diagram and typical spectra at different points in the receiver are depicted in Fig. 6.3b. The symbols have a clear and distinct shape.

To quantify the quality of the received signals, we performed BER estimations [133]. The EVM measurements were found to be between 9 % and 15.8 %. Estimations [133] and the results (Fig. 6.4b) discussed in the methods section indicate that the BER of all 75 subcarriers are below the 1.9×10^{-2} third generation forward error correction (FEC) limit [134] with less than 25 % overhead, but partially exceeding the 2.3×10^{-3} FEC limit with 7 % overhead. This is slightly within limits considered to allow error-free transmission according to third generation FEC standards [134]. The achieved line rate of 10.8 Tbit s^{-1} thus corresponds to a net error-free data rate of 8.64 Tbit s^{-1}. The quality of the 16QAM transmitter and receiver are limited by the currently available electronic bandwidths – and not by the optical concept [10].

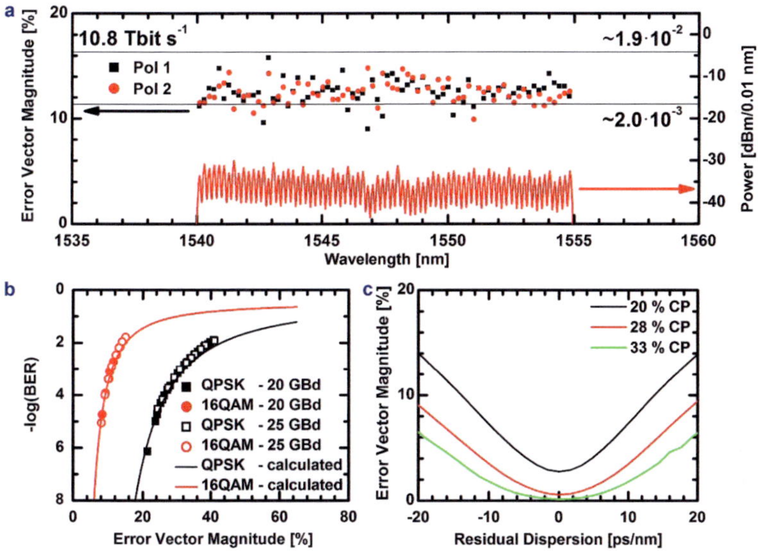

Fig. 6.4: All-optical OFDM results. **a** Measured error-vector magnitude (EVM) for both polarizations (■, ●) and for all subcarriers of the OFDM signal decoded with the all-optical FFT. The estimated BER for all subcarriers is below the third-generation FEC limit of $1.9\times10{-2}$. The optical spectrum (——, right-hand side vertical axis) is drawn beneath. **b** Relation between BER and EVM. Measured points (dots) and calculated curves [132] (lines) of BER as a function of EVM for QPSK and 16QAM. **c** Tolerance towards residual chromatic dispersion of the implemented system decoded with the 8-point FFT for a cyclic prefix $CP = \tau_c / (T_s + \tau_c)$ of 20 %, 28 % and 33 %.

6.1.2 Transmission Performance of All–Optical OFDM Signals

The large bandwidth of the OFDM signal has implications for the transmission. On the one hand, there will be dispersion within each subcarrier and on the other hand, there will be walk-off between subcarriers. Due to the subcarrier spacing of several GHz it makes sense to transmit the signal over a dispersion compensated link. However, in contrast to OTDM, our OFDM signal can tolerate quite high amounts of residual dispersion, and higher order dispersion compensation is not necessary. The achievable transmission distance of such an OFDM signal is only limited by the accumulated dispersion within the optical bandwidth of the subcarrier receiver. It can therefore tolerate quite some residual dispersion. To support this we include some comprehensive computer simulations (Fig. 6.4c) as outlined in the methods section. For instance, for a transmission link within a standard single mode fibre (SMF) with a dispersion of 17 ps/(nm×km) and standard dispersion compensation modules, typical residual dispersion as low as 0.01 ps/(km×nm) can be achieved [135]. After transmitting the signal e.g. over 50 km the residual dispersion therefore would be in the order of 0.5 ps, which would not distort the signal significantly (Fig. 6.4c).

6.1.3 Receiver Concept Comparison

To judge the effectiveness of our FFT receiver, we also implemented three alternative receiver concepts and tested them for a QPSK signal. A QPSK signal was chosen as it was not possible to receive a 16QAM signal with the alternate receivers due to their inferior performance. First, we extracted a subcarrier with a narrow bandpass filter (XTRACT, Anritsu) that was adjusted in bandwidth for best performance of the received signal, Fig. 6.5a. The selected filter bandwidth was approximately 25 GHz. The constellation diagram showed significant distortions. When using narrow optical filtering, one has to accept a compromise between crosstalk from neighbouring channels as modulated OFDM subcarriers necessarily overlap, and intersymbol interference due to the increasing length of the impulse response when narrow filters are used. Narrow filters can be used, though, if the ringing from intersymbol interference is mitigated by an additional time gating, see Fig. 6.5b. We then tested reception of a subcarrier using a coherent receiver only, Fig. 6.5c. In the coherent receiver, the signal is down-converted in a 90°-hybrid and detected using balanced detectors and a real-time oscilloscope. Using a combination of electrical low-pass filtering due to the limited electrical bandwidth of the oscilloscope and digital signal processing, the subcarrier is then extracted from the received signal. This receiver performs better than the filtering approach, but a larger electrical bandwidth and sampling rate of the analogue-to-digital converters and additional digital signal processing would be needed to get rid of crosstalk from other subcarriers and to achieve a performance similar to the optical FFT [38]. Finally, we compared the receivers from Fig. 6.5a, b and c with results from the implementation of our all-optical 8-point FFT receiver in Fig. 6.5d. From the constellation diagrams in Fig. 6.5a-d it can be seen that the all-optical FFT clearly outperforms the other solutions.

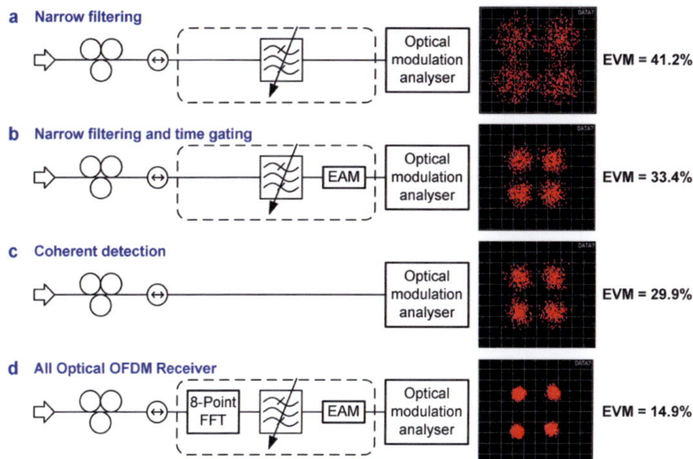

Fig. 6.5: Experimental comparison of receiver concepts. **a** Detection using spectrally narrow filtering with an optical bandpass filter (XTRACT Anritsu). The filter bandwidth is adjusted for optimum performance, **b** detection using the narrow filter with an additional gate, **c** coherent detection, where the signal is directly down-converted in an optical 90°-hybrid and afterwards filtered electronically, and **d** detection with the all-optical FFT.

6.1.4 System Demonstration at a line rate of 26 Tbit s^{-1}

Finally, we demonstrate the scalability and viability of the all-optical processing concept for a 26 Tbit s^{-1} OFDM signal (Fig. 6.6), where we additionally transmit the signal derived from a single source over a distance of 50 km. In contrast to the 10.8 Tbit s^{-1} experiment we improve the comb generator by employing a spectral slicing technique to enhance the comb quality. In the comb generator, the output of a 12.5 GHz MLL (Ergo-XG) is amplified, filtered (Fig. 6.6a, spectrum 1) and split into two paths. One path is spectrally broadened in a highly nonlinear photonic crystal fibre (Fig. 6.6a, spectrum 2). The broadened spectrum is then equalized, and the low-power, unstable part in the centre is replaced by the original spectrum from the second path (Fig. 6.6 spectrum 3). This results in a flat and stable comb spectrum with 336 subcarriers. The spectrum is then split into odd and even subcarriers using an optical dis-interleaver (DIL) with a neighbour channel suppression of more than 35 dB. Odd and even subcarriers are modulated independently employing two 16QAM transmitters at 10 GBd. With the carrier spacing of 12.5 GHz and a symbol duration of 80 ps this results in a cyclic prefix length of 20 ps (CP = 20 %). After recombining the subcarriers, polarisation multiplexing is emulated in analogy to the 10.8 Tbit s^{-1} experiment.

The OFDM signal with an average power of 9.5 dBm is then launched into a single span of single-mode fibre SMF-28 having a length of 50 km and a span loss of 10.17 dB. After transmission, the signal is amplified and the dispersion of the link is compensated in a dispersion compensation module with a total insertion loss of 4.8 dB. The residual dispersion of the link is 17 ps/nm.

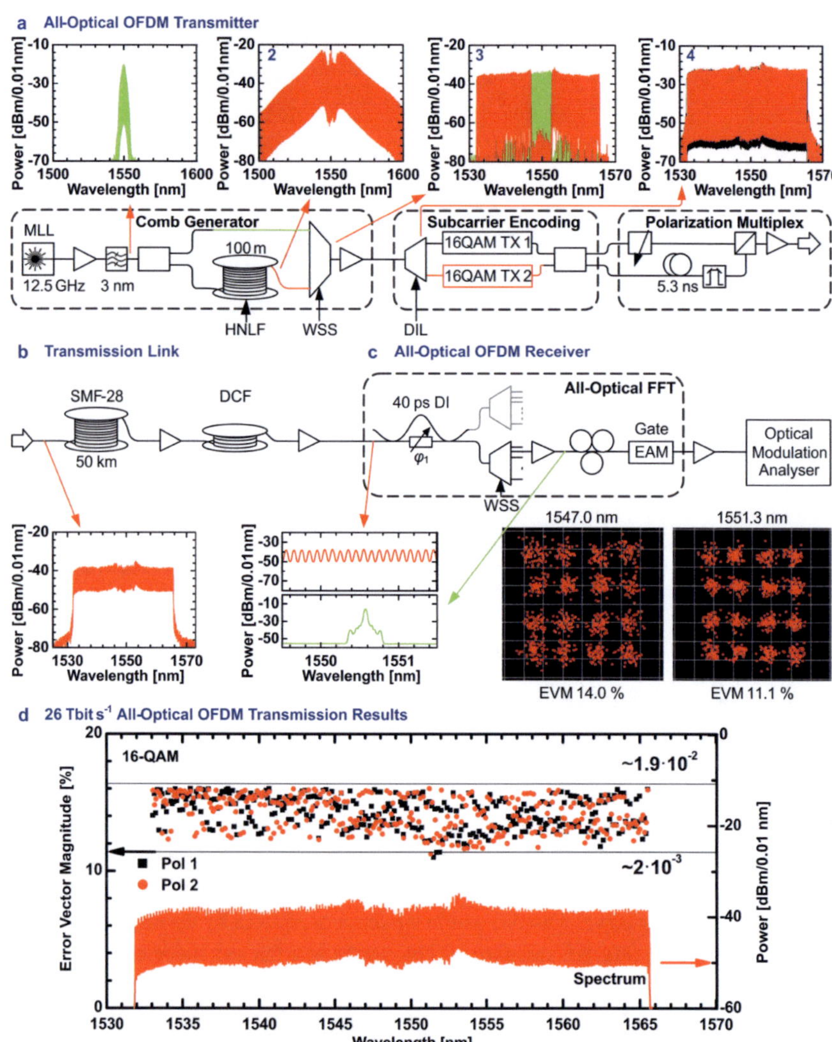

Fig. 6.6: All-optical 26 Tbit s−1 OFDM transmission experiment. **a** All-optical OFDM transmitter with advanced comb generator employing spectral slicing to generate 325 carriers. **b** Transmission link with 50 km of standard single mode fiber (SMF-28) and C (DCF). **c** All-optical OFDM receiver. **d** Error vector magnitudes for both polarizations (■, ●) and for all 325 OFDM subcarriers after transmission over 50 km of dispersion compensated SMF.

The receiver comprises an all-optical FFT circuit and an optical modulation analyser as with the 10.8 Tbit s^{-1} experiment to perform coherent detection and EVM analysis. Here we implement the optical FFT using a single delay interferometer realizing a two-point optical FFT [10]. This was preferred over the 8-point FFT used in the 10.8 Gbit s^{-1} demonstration due its simplicity. The penalty for this simplification has been shown to be within limits [10]. A wavelength-selective switch (Finisar Waveshaper) isolated the desired frequency band. The time gating was performed by an electro absorption modulator (EAM – CIP 40G-PS-EAM-1550) at 10 GHz. In this demonstration, the phase between data signal and electrical clock is adjusted manually, which can be avoided using an electronic clock recovery to extract the clock phase of the incoming OFDM signal [136]. The subcarriers are received and demodulated by an OMA (Agilent N4391A) with an electrical bandwidth of 16 GHz (Agilent Infiniium DSOX93204A). The signal is equalized with a 21 tap equalizer. For each polarization, the error vector magnitude is calculated and averaged over 10 received sequences of 1,024 symbols. This measurement was carried out for 325 subcarriers of the OFDM signal – the 11 subcarriers with the shortest wavelength were not measured as the subcarrier quality was severely degraded due to their inferior optical signal-to-noise ratio.

The comb generator produced a stable frequency comb from 1533 nm to 1565 nm with a flatness of < 0.5 dB. The evaluated EVM is shown in Fig. 6.6d for all measured 325 subcarriers. Typical constellation diagrams are depicted in Fig. 6.6c. Similar to the 10.8 Tbit s^{-1} experiment we performed a BER estimation [133] – see methods section. It can be seen that the EVM for all subcarriers after transmission are between 11 % and 16 % with an average EVM of 14.2 %. The estimations indicate that all 325 subcarriers are below the BER = 1.9×10^{-2} third generation FEC limit with 25 % overhead. The demonstrated line rate of 26 Tbit s^{-1} therefore corresponds to a net error-free data rate of 20.8 Tbit s^{-1}. Degradation of the EVM due to transmission is negligible as predicted in the previous section. The achieved spectral efficiency for the modulated subcarriers is 6.3 bit/s/Hz.

6.1.4.1 Tolerance to Residual Dispersion

To determine the tolerance of such an OFDM signal to residual dispersion we performed comprehensive computer simulations. The general setup is that of Fig. 6.3, except that we did not use polarization multiplexing in the simulations. In the simulation we use 64 subcarriers with a spacing Δf of 25 GHz. This corresponds to an OFDM symbol length T_S of 40 ps without cyclic prefix. We do simulations for a number of 512 OFDM symbols and cyclic prefix durations T_{CP} of 10, 15.5 and 20 ps. The electrical bandwidth of the transmitter is 25 GHz and the receiver uses an 8-point-FFT, an optical receiver filter (2nd order Gaussian, $8 \times \Delta f = 200$ GHz bandwidth) and an EAM gate with a FWHM of 4 ps. The electrical bandwidth of the receiver is $0.7 \times \Delta f = 17.5$ GHz. In the transmitter, the subcarriers are separately but synchronously QPSK modulated and combined into one OFDM channel. A residual dispersion between −60 ps/nm and +60 ps/nm is added to the signal to test its tolerance to residual dispersion, as this determines how well the dispersion of the link must be compensated. At the receiver, a 200 GHz OFDM subband is extracted using an optical filter with a 2nd-order Gaussian transfer function. The optical FFT is performed on this subband, and the subcarrier located at the centre of the filter passband is analysed in terms of the EVM. To show the impact of a cyclic prefix we did the simulation with a cyclic prefix $CP = \tau_c / (T_s + \tau_c)$ of 20 %, 28% and 33 % which corresponds to symbol rates of 20 GBd, 18 GBd and 16.6 GBd, respectively. The percentage of the cyclic prefix is calculated with respect to the associated reciprocal symbol rates 50 ps, 55.5 ps and 60.2 ps. The bandwidth limitations of the system were included to obtain realistic results as shown in Fig. 6.4c. Additional noise was not taken into account. First, it can be seen that a certain cyclic prefix is required due to the bandwidth limitations of the system. Second, in our system with a symbol rate of 18 GBd and CP = 28 %, a residual dispersion of $C_{res,max} = \pm 5\,ps\,/\,nm$ can be tolerated with an EVM penalty lower than 1 %. It should be noted, that a state-of-the art combination of normal and dispersion compensating fibres allows for a residual dispersion variation of less than $C_{res} = 0.01\,ps\,/\,(km\,nm)$ within a bandwidth of 20 nm [135]. This corresponds to an achievable transmission distance of $d_{max} = \left| C_{res,max} \right| / C_{res} = 500\,km$ with an EVM penalty of less than 1 %.

6.1.5 Summary

In summary, we demonstrated for the first time the real-time generation and real-time FFT-processing of 10.8 Tbit s^{-1} and 26 Tbit s^{-1} line-rate OFDM signals employing an all-optical FFT scheme based on cascaded delay interferometers and a time gate. We extracted the OFDM subcarriers sequentially. Additionally, we demonstrate transmission of a 26 Tbit s^{-1} line-rate OFDM signal over 50 km of dispersion-compensated fibre. For the first time we realized signal reception at 26 Tbit s^{-1} using the optical FFT. This technology enables reception of ultra-fast OFDM signals with state-of-the-art electronics at speeds far beyond the limits of electronics, avoiding the electronic bottleneck.

6.2 Nyquist WDM record experiment

Single–laser 32.5 Tbit/s Nyquist WDM transmission

D. Hillerkuss, R. Schmogrow, M. Meyer, S. Wolf, M. Jordan, P. Kleinow, N. Lindenmann, P. Schindler, A. Melikyan, X. Yang, S. Ben-Ezra, B. Nebendahl, M. Dreschmann, J. Meyer, F. Parmigiani, P. Petropoulos, B. Resan, A. Oehler, K. Weingarten, L. Altenhain, T. Ellermeyer, M. Moeller, M. Huebner, J. Becker, C. Koos, W. Freude, and J. Leuthold.

This paper has been published as [12]. First draft published in ArXiv e-prints [13], also published in [14] and [15].

We demonstrate single laser 32.5 Tbit/s 16QAM Nyquist WDM transmission over a total length of 227 km of SMF-28 without optical dispersion compensation. A number of 325 optical carriers are derived from a single laser and encoded with dual-polarization 16QAM data using sinc-shaped Nyquist pulses. As we use no guard bands, the carriers have a spacing of 12.5 GHz equal to the symbol rate or Nyquist bandwidth of the data. We achieve a net spectral efficiency of 6.4 bit/s/Hz using a software-defined transmitter, which generates the electric drive-signals for the electro-optic modulator in real-time.

6.2.1 Introduction

Super-channels for multi-Tbit/s transmission are envisioned to play an important role in future optical networks [137]. Such channels typically consist of one carrier or several frequency-locked carriers onto which data are encoded [8, 33, 43, 124]. As cost and power consumption are important issues [138-140], a reduced component count is desirable. Therefore, single-laser Tbit/s transmission systems are of special interest.

Up to the year 2009, single-laser Tbit/s systems were mostly implemented using optical time division multiplexing (OTDM) [33, 124]. With OTDM, data rates of up to 10.2 Tbit/s were obtained within an optical bandwidth of roughly 30 nm or 3.75 THz. This corresponds to a net spectral efficiency of 2.6 bit/s/Hz [33]. With TDM, transmission at 10.2 Tbit/s over 29 km of dispersion-managed fiber has been demonstrated.

Since 2005, multicarrier transmission has attracted increasing interest as it offers highest spectral efficiency. In particular, coherent wavelength division multiplexing (CO-WDM) [45] and orthogonal frequency division multiplexing (OFDM) [31, 37, 127] have been proposed. The first demonstration of an OFDM-signal beyond 1.0 Tbit/s in 2009 showed a transmission distance of 600 km using standard single mode fiber [40]. The spectral efficiency was 3.3 bit/s/Hz. In 2010, using an optical FFT scheme [10], we were able to encode and detect a 10.8 Tbit/s super-channel [7]. Subsequently, we generated and transmitted an OFDM super-channel with a line rate of 26 Tbit/s over a distance of 50 km of standard single mode fiber with standard dispersion compensating modules [8]. The net spectral efficiency could be increased to 5.0 bit/s/Hz.

Nyquist pulse shaping [67] is an alternate method to improve the transmission performance. Such a pulse shaping can increase the nonlinear impairment tolerance [69, 141, 142], reduce the spectral footprint of single-channel signals [67], and therefore reduce both receiver complexity [143] and the required channel spacing in WDM systems [41, 144]. Of particular interest are sinc-shaped Nyquist pulses, which have a rectangular spectrum [5, 6, 67] and confine the signal to its Nyquist bandwidth [67]. This enables highest intra-channel spectral efficiencies, and has recently enabled transmission with a net spectral efficiency of 15 bit/s/Hz [145]. Combining a number of adjacent spectra, we end up with Nyquist WDM, where the carrier spacing is equal to the symbol rate (assuming identical symbol rates in all bands, which is not necessarily required). In contrast to CO-WDM where the phase of neighboring subcarriers is adjusted to minimize crosstalk [45], phase control of the carriers is not necessary for Nyquist WDM and OFDM [8]. Recently, this has been discussed as an option for Tbit/s super-channels [41], and favorable transmission properties have been predicted based on the fact that a train of modulated sinc-shaped Nyquist pulses has a relatively low peak-to-average power ratio [6]. Subsequently, several successful experiments demonstrated WDM with Nyquist pulse shaping and small guard bands [43, 44]. Also, a first demonstration of Nyquist WDM at 400 Gbit/s with a net spectral efficiency of 3.7 bit/s/Hz has been shown [42]. An arbitrary waveform generator (AWG) was used to create a Nyquist signal that was computed offline using a 601-tap finite impulse response (FIR) filter.

Nyquist WDM transmission with real-time sinc-pulse shaping and 16QAM has not yet been shown. The problem lies in the limited-length representation of acausal sinc-pulses in systems with real-time signal processing, where a practicable number of FIR filter taps has to be used. So far it had not been clear if a real-time computation will support Nyquist WDM transmission over significant distances.

In this paper, we report single-laser Nyquist WDM super-channel transmission at a record high aggregate line rate of 32.5 Tbit/s. This is the largest aggregate line rate ever encoded onto a single laser. In contrast to other experiments [42-44], we use neither offline processing at the transmitter nor guard bands. The electrical Nyquist signals are computed in real-time using a 64-tap FIR filter. The net spectral efficiency is 6.4 bit/s/Hz. We show transmission over 227 km. The achieved transmission distance is more than four times longer than what was reported for the most recent OFDM super-channel experiment [8]. The present experiment further shows how frequency comb generation is maturing. Here, 325 frequency locked carriers are generated from one source. This enables Tbit/s Nyquist WDM transmission with sufficient OSNR to reach distances of several hundred kilometers.

6.2.2 Benefits and challenges of Nyquist WDM transmission sys-

tems

There are significant advantages and challenges in the implementation of Nyquist WDM as compared to other schemes like standard WDM, OFDM and all-optical OFDM transmission systems. True Nyquist WDM builds on immediately neighbored partial spectra that do not overlap due to their rectangular shape. The bandwidth (B_N) of the partial spectra of channel N is given by the Nyquist bandwidth of the encoded data, and it is equal to the symbol rate R_N. To achieve the rectangular partial spectrum, sinc-shaped Nyquist pulses are needed.

Due to the minimized bandwidth, Nyquist pulse transmitters have substantial benefits compared to standard non-return-to-zero transmitters (NRZ), because the available electrical bandwidth of components like digital-to-analog converters (DAC), driver amplifiers and modulators is optimally used. This, however, comes at the price of an increased amount of digital signal processing. Also, some oversampling is needed to accommodate anti-aliasing filters after the digital to analog converters [6].

A challenge when generating true Nyquist WDM is that the distance of the neighboring optical carriers has to be controlled precisely. For a true Nyquist WDM signal, the following condition has to be met: If two neighboring channels, namely channel N and channel $N+1$, operate at symbol rates R_N and R_{N+1}, respectively, the spacing Δf has to be $\Delta f = (R_N + R_{N+1})/2$. If the spacing is larger, optical bandwidth is wasted. If the spacing is smaller, linear crosstalk from neighboring Nyquist channels will increase significantly. In our experiment we investigate the case, where all carriers have the same symbol rate R. In this case, the carrier spacing Δf is equal to the symbol rate R.

Nyquist WDM transmission, when compared to all-optical OFDM [8], has the distinct advantage that only the dispersion within the bandwidth (B_N) of one Nyquist channel has to be compensated. Usually this is not done with dispersion compensating fibers, but rather by digital signal processing in the coherent receiver as it is the case for the present experiments. Electronic dispersion compensation can also be implemented for individual or small groups of subcarriers of an OFDM signal. Because the smallest optical filter bandwidth is limited by technical constraints, the all-optical OFDM symbol rate is in the same order of magnitude as the symbol rate of the Nyquist WDM channels. As the bandwidth of each modulated OFDM subcarrier is significantly larger than the OFDM symbol rate R_N, a larger amount of digital signal processing for dispersion compensation is required when compared to Nyquist WDM. Additionally, simulations in [6] indicate that Nyquist pulse shaped signals exhibit a lower peak to average power ratio (PAPR) compared to OFDM.

In Nyquist WDM receivers, the WDM channels are coarsely selected with optical filters. The final selection of a channel is done in the electrical domain by digital signal processing whereby sharp-edged filters can be realized. An additional challenge arises when implementing a clock recovery for Nyquist signals. As they will not have noticeable spectral components at the frequency of the sampling clock, most standard clock recovery algorithms will not work. Our solution for a clock recovery has been presented in [6].

6.2.3 Nyquist WDM System Concept

The envisioned Nyquist WDM super-channel concept with transmitter and receiver is illustrated in Fig. 6.7. The transmitter can be separated into two main parts: the carrier generation and the carrier modulation.

The carrier generation is a key part of this scheme as a small frequency drift of the optical carriers will immediately lead to an increased crosstalk from and to neighboring Nyquist WDM channels. It therefore makes sense to use an optical comb source where the generated optical carriers are inherently equidistant in frequency. This comb source could be replaced by 325 precisely-stabilized narrow line-width lasers. However, as the comb source only requires a single mode-locked laser, two EDFAs, a highly nonlinear fiber (HNLF) and a waveshaper (WS), it is doubtful that 325 lasers with additional frequency stabilization circuits could be operated with similar energy efficiency. To separate these carriers, we propose to use a cascade of an optical interleaver (IL) and two optical wavelength demultiplexers (DMX).

For carrier modulation, we propose the use of Nyquist transmitters with digital signal processing (DSP), as these transmitters can generate signal spectra with an extremely steep roll off [5, 6]. To maintain the rectangular shape of the Nyquist spectra, special care has to be taken when combining different Nyquist channels. To this end, we select a combination of two optical multiplexers and an optical coupler. The two optical multiplexers combine odd or even channels, respectively, without limiting the Nyquist spectra. The optical coupler combines odd and even channels for transmission over the optical fiber link. The transmitter scheme is depicted on the left hand side of Fig. 6.7.

In the Nyquist WDM receiver, the signal is split by a coupler, and two optical demultiplexers separate odd or even optical carriers in front of the Nyquist receivers. Even though it would be possible to simply split the signal and receive it using multiple coherent Nyquist receivers, we suggest including optical demultiplexers for two reasons: First, this reduces the insertion loss of the receiver structure significantly. Second, as the total optical power of the signal in a coherent receiver is restricted by physical limitations of the photodiodes, the power of the received signal can be increased significantly, and unnecessary power loading of the balanced detectors with a large number of unwanted carriers is avoided. After filtering, residual components of neighboring Nyquist channels remain (see bottom right inset in Fig. 6.7). These residual spectra have to be removed by digital brick wall filtering in the Nyquist receivers.

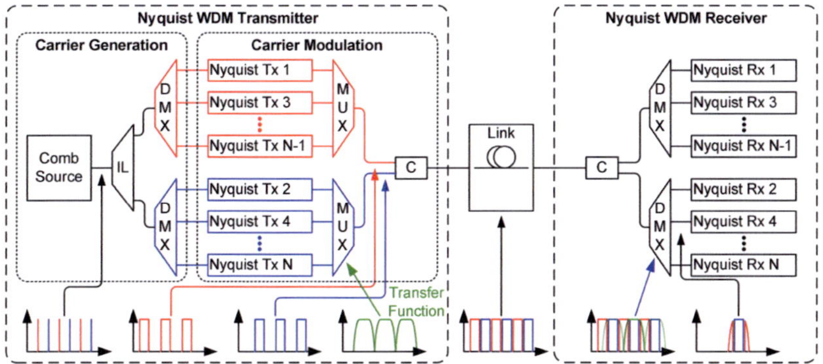

Fig. 6.7: Concept for a Nyquist WDM transmission system. In the transmitter, an optical comb source generates the optical carriers. Filters, namely an interleaver (IL) and an optical demultiplexer (DMX), separate the optical carriers. Nyquist transmitters (Nyquist TX1, 2, … N) encode the data. Multiplexers (MUX) and a standard optical coupler (C) combine the transmitter outputs. After transmission, a coupler (C) and optical demultiplexers (DMX) split the carriers for the Nyquist receivers (Nyquist RX1, 2, … N). The schematic spectra illustrate the Nyquist WDM transmitter and receiver concept for a total number of $N = 6$ optical carriers. Spectra of odd carriers (——), spectra of even carriers (——), and MUX/DMX transfer functions (——) are shown.

6.2.4 Implementation of a Nyquist pulse transmitter

Pulse shaping is crucial for the implementation of Nyquist WDM transmission systems. Sinc-shaped Nyquist pulses extend infinitely in time and generate a rectangular spectrum (insets in Fig. 6.8). The Nyquist pulses repeat with the impulse spacing T, are modulated with complex data, and have a total bandwidth $B = 1 / T$, which equals the symbol rate R. For our experiment, two real-time Nyquist pulse transmitters are implemented (one is shown in Fig. 6.8) to modulate odd and even carriers. The transmitter setup is based on the setup in [5, 6], which was developed from our multiformat transmitter [3]. For an efficient pulse shaping, we use

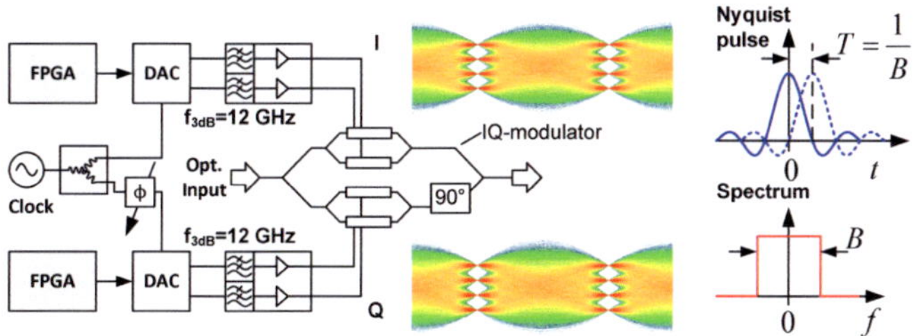

Fig. 6.8: Implementation of real-time transmitter for sinc-shaped Nyquist pulses. Two FPGAs are programmed to generate the PRBS (2^{15}-1), to perform symbol mapping, and to provide pulse shaping with an FIR-filter. The in-phase (I) and quadrature (Q) signals are converted into the analog domain using two high speed VEGA-DACs. Image spectra are removed with additional anti-aliasing filters, and the resulting signals are amplified for driving the optical IQ-modulator. Insets: Simulated eye diagrams for I and Q drive signals, sinc-pulse and corresponding spectrum at the output of the transmitter. The simulated eye diagrams illustrate that there is only a small tolerance in clock phase when receiving such a signal, which makes a proper clock phase recovery procedure extremely important [23].

oversampling with two samples per symbol. In a first step, a pseudo-random bit sequence (PRBS, length $2^{15}-1$) is generated in real-time by the two FPGAs (Xilinx XCV5FX200T). The 16QAM symbols enter a FIR filter with 64 taps, thereby generating sinc-shaped Nyquist pulses modulated with the data. The number of taps directly impacts the processing delay of the transmitter and is limited by the available space for logic on the FPGA. The resulting digital signal is then converted to the analog domain using either two VEGA DAC25 (Tx1 in Fig. 6.9 (a)), or two VEGA DAC-II (Tx2 inFig. 6.9 (a)). We use different DACs for the transmitters due to their availability in our laboratory. We did not observe a significant performance difference of the two transmitters. Simulated eye diagrams are displayed as insets in Fig. 6.8. The DACs operate at 25 GSa/s, generating 12.5 GBd signals with an electrical bandwidth of 6.25 GHz. In contrast to [5, 6], we use electrical lowpass filters with a 3 dB bandwidth of 12 GHz and a suppression of > 30 dB at 13 GHz to remove the image spectra. After amplification, the signals are fed to an optical IQ-modulator which in turn modulates the optical carrier [6].

6.2.5 Nyquist WDM Experiment

Our Nyquist WDM system Fig. 6.9 (a)–(c) consists of four main components: the optical comb source, two Nyquist WDM transmitters as in Fig. 6.8, a polarization multiplexing emulator, and a coherent optical Nyquist WDM receiver.

The optical comb source is actually one of the key components in this experiment. It did not only provide a cost-effective and energy-efficient way to generate a large number of optical carriers, it also generated them with a highly stable frequency spacing, which is useful for the case of equal symbol rates in all channels having a carrier spacing equal to the symbol rate. This comb source uses a pulse train from an ERGO-XG mode-locked laser (MLL) which is amplified and filtered to remove amplified spontaneous emission. The MLL output is split in two parts, one of which is spectrally broadened in a highly nonlinear photonic crystal fiber [8]. In the waveshaper (WS), the original MLL comb is bandpass-filtered and fills the void in the notch-filtered broadened comb such that unstable sections in the center of the broadened spectrum are replaced by the original MLL spectrum. This spectral composing process is also exploited for equalizing the frequency comb to form a flat output spectrum. For Nyquist WDM transmission there is no requirement for a stabilization scheme to guarantee a fixed initial phase of all carriers relative to the beginning of a symbol time slot (as is the case for coherent WDM [45]). A number of 325 optical carriers are generated between 1533.47 and 1566.22 nm with a spacing of 12.5 GHz. Our measurements indicate that the line width of the carriers is significantly lower than the line width of our local oscillator in the receiver (Agilent 81682A – external cavity laser – ECL – line width typ. 100 kHz). The MLL is adjusted such that the carriers fall on the ITU grid. This allows us to use off-the-shelf optical components.

For modulation, the spectral lines are decomposed into odd and even carriers using a standard optical interleaver. Odd and even carriers are modulated with Nyquist transmitters

Tx1 and Tx2, respectively, see Fig. 6.9. To generate a Nyquist WDM signal, the symbol rate of 12.5 GBd is chosen to equal the carrier spacing of 12.5 GHz. Both transmitters operate with separate sampling clock sources as no symbol synchronization is required. After modulation, odd and even carriers are combined in an optical coupler to form the Nyquist WDM signal. The two outputs of this coupler are then delayed relative to each other for data decorrelation (delay 5.3 ns) and combined in a polarization beam combiner to emulate polarization multiplexing.

The signal is then amplified, and transmitted over distances of 75.78 km and 227.34 km using a Corning SMF-28 with EDFA-only amplification. The optimum launch power was found to be 18 dBm for the complete Nyquist WDM signal. This corresponds to a power of -7 dBm per carrier. This launch power was optimized for the carrier at 1550.015 nm.

After transmission, the carrier of interest is selected in a WS, amplified, and finally received in an optical modulation analyzer (OMA – Agilent N4391A). Four signal processing steps were performed before the 16QAM demodulation. First, the chromatic dispersion is compensated. Second, a digital brick wall filter selects one channel and removes all remainders of neighboring channels. Third, the standard polarization tracking algorithm [146] separates the two polarizations, and fourth, the clock phase is recovered as described in [6]. Only the digital brick wall filtering and the clock phase estimation algorithm had to be implemented in addition to the standard algorithms included in the OMA software. No modification of the 16QAM receiver algorithms of the OMA software was required. Using the built-in frequency offset estimation algorithm, we were able to precisely determine the frequency offset up to 500 MHz. This allows for frequency tracking without requiring pilot symbols, which would lead to additional overhead. A least mean square adapted linear FIR filter with 51 taps was used as equalizer to compensate for the frequency dependence of the overall transmission system. PMD was not compensated for.

To characterize the signal quality, we measured the error vector magnitude (EVM). We derived a bit error ratio (BER) estimate from the EVM data [133]. We verified this estimate for selected points to support the applicability of this estimation technique. The BER was measured with the OMA for carriers in the back-to-back case and for the transmission over 227.34 km. We chose carriers that exhibited higher EVM values, which enabled us to provide an upper limit of the BER. The accuracy of the BER estimation method from [133] has been demonstrated experimentally in [85, 147] and was again confirmed in these experiments.

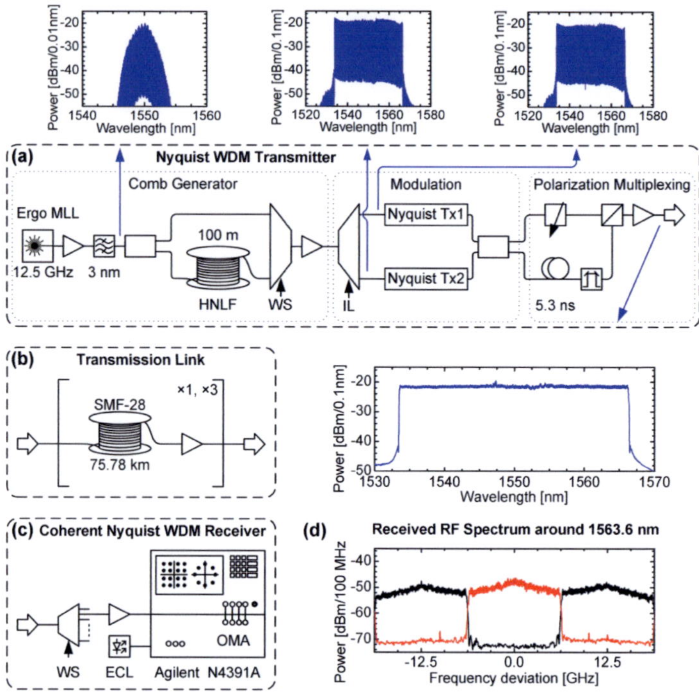

Fig. 6.9. Nyquist WDM setup. (a) A mode-locked laser (MLL) produces a frequency comb (see inset above (a)) which is broadened in a highly nonlinear fiber (HNLF). A waveshaper (WS) equalizes the resulting comb of 325 lines and replaces the unstable central part by a copy of the original MLL comb. An optical interleaver (IL) separates odd and even carriers (power spectra in the inset above (a)). On these carriers the transmitters Tx1 and Tx2 (see schematic in Fig. 6.8) encode 16QAM data in form of sinc-shaped Nyquist pulses. Polarization multiplexing is emulated. The arrow points to the resulting optical power spectrum (inset below (a)). (b) The optical signal is then transmitted over one or up to three spans of Corning SMF-28 with EDFA-only amplification. (c) In a coherent Nyquist WDM receiver a WS selects a 60 GHz wide group of Nyquist channels fitting to the bandwidth of the optical modulation analyzer (OMA). An external cavity laser (ECL – Agilent 81682A) provides the local oscillator for the coherent receiver. (d) Two-sided RF power spectrum after down conversion from an optical carrier at 1563.6 nm (——) and from the adjacent carriers (——). The rectangular shape of the spectra proves the effectiveness of the Nyquist pulse shaping.

6.2.6 Experimental Results

Back-to-back measurements serve as a reference for the overall system performance, Fig. 6.10 (a). The EVM for almost all carriers was below the threshold for second generation FEC (BER = 2.3×10^{-3}). The total EVM (defined as the root mean square of the EVM of all carriers) was $\mathrm{EVM_{tot}} = 10.3$ %. In Fig. 6.10 (b, c) we show the results after transmission. $\mathrm{EVM_{tot}}$ degrades by 1.0 percentage points for a distance of 75.78 km and by 1.7 percentage points for a distance of 227.34 km. The EVM for all carriers and distances is well below the limiting EVM for a BER of 1.8×10^{-2} using next generation soft decision FEC [134]. The

EVM differences between the two polarizations, especially in the outer wavelength range, are due to the wavelength dependence of the 3 dB coupler in the polarization multiplexing scheme. An additional degradation can be traced back to the non-ideally gain-flattened ED-FAs that were available for the experiment. These EDFAs lead to the uneven received spectra after transmission. If EDFAs with better gain flattening were used in such an experiment, we would expect a significant increase in achievable transmission distance.

When approaching the Nyquist spacing of the channels, a certain amount of linear crosstalk is to be expected due to the finite filter slopes. The finite impulse response (FIR) filters are implemented by digital real-time signal processing [5, 6]. In the present experiment, the crosstalk is very small as shown in the RF spectra in Fig. 6.9 (d). Here we show the measured RF spectra when only transmitter 1 (——) or transmitter 2 (——) are turned on. A small amount of sampling clock leakage in the DAC can be observed at 12.5 GHz. The tones around ±10 GHz originate from the sampling oscilloscope in the receiver. The EVM degradation due to residual crosstalk is measured to be 1.5 to 2 percentage points. To quantify the influence of linear crosstalk from neighboring Nyquist channels, we measured the BER for the wavelengths 1563.66 nm and 1564.58 nm with and without neighboring channels. For 1563.66 nm, the BER increased from 6.3×10^{-4} without neighboring channels to 1.2×10^{-3} with neighboring channels. For 1564.58 nm the BER increased from 1.4×10^{-4} to 2.1×10^{-3}. This shows that the back-to-back performance is mainly limited by linear crosstalk due to the limited number of taps. This crosstalk could be reduced by increasing the carrier spacing from the Nyquist case to a larger spacing. This could increase the achievable transmission distance or reduce the required amount of FEC overhead — however, such a system would no longer be a Nyquist WDM system. The line rate of 32.5 Tbit/s corresponds to a net data rate of 26 Tbit/s for the transmitted signals (taking into account the 25 % FEC overhead [134]).

To verify the relationship between EVM and BER, we chose to measure BER and EVM for some selected carriers. Due to time constraints, we measured the BER only for the carriers presented in Tab. 6.1. We chose these carriers such that the worst-case EVM for the back-to-back and 227 km transmission was included. In Tab. 6.1 we show the measured EVM values and the measured BER. In addition, we calculated a BER corresponding to the measured EVM values using Eq. (5.17), which was derived in reference [133]::

$$\text{BER} \approx \frac{1-M^{-\frac{1}{2}}}{\frac{1}{2}\log_2 M}\, \text{erfc}\left[\sqrt{\frac{3/2}{(M-1)(k\,\text{EVM})^2}}\right] . \tag{5.17}$$

For our 16QAM signal the number of all possible constellation points is $M = 16$, the number of bits encoded in one QAM symbol is $\log_2 M = 4$, and the modulation format dependent factor is $k^2 = 9/5$. It has to be mentioned that a misprint [147] happened in equation (4) in reference [85], where $\sqrt{2}$ has to be replaced by 1. However, the corresponding plots in Fig. 3b of reference [85] were calculated correctly [147]. The calculated BER differs only slightly from the measured BER, supporting the applicability of the BER-EVM relationship, which is based on a Gaussian hypothesis.

The distribution of the noise on the constellation points is a critical factor for the applicability of the BER-EVM relationship, and for predicting error-free operation when using soft-decision FEC [134]. To support the claim that our constellation field vectors are perturbed by additive Gaussian noise, we plotted in Fig. 6.11 the constellation diagrams for the carriers listed in Tab. 6.1. Additionally, we performed a statistical analysis of the distribution of in-phase (I) and quadrature-phase (Q) error vectors in the constellation diagram. The results are displayed in Fig. 6.11 and support the claim of a Gaussian distribution of the added noise.

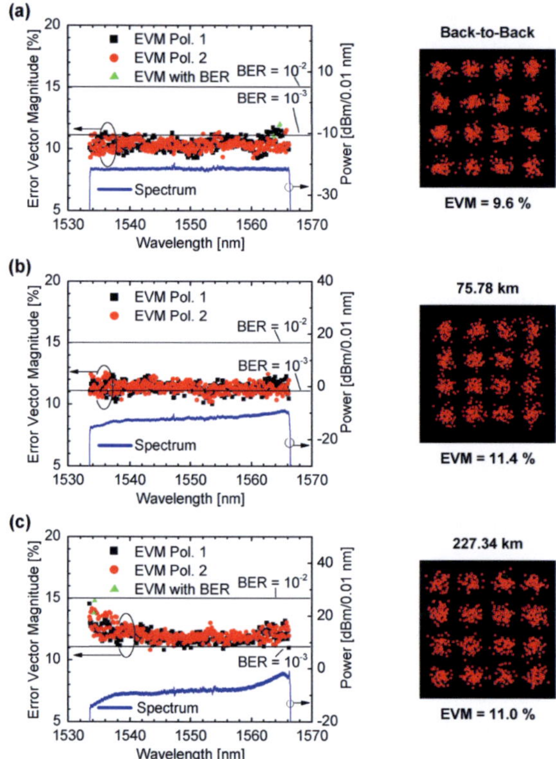

Fig. 6.10: Experimental results for the transmission experiment. The measured error vector magnitudes for all subcarriers (■ – polarization 1, ● – polarization 2) and the received optical spectra (——) are plotted for (a) back-to-back characterization, (b) transmission over 75.78 km, and (c) transmission over 227.34 km. For selected EVM values in the area around the highest EVM values (▲), the BER was measured and the results are displayed in Tab. 6.1 to verify the applicability of BER estimation from EVM measurements. Corresponding constellation diagrams for the carrier at 1550.015 nm are shown for the various transmission distances.

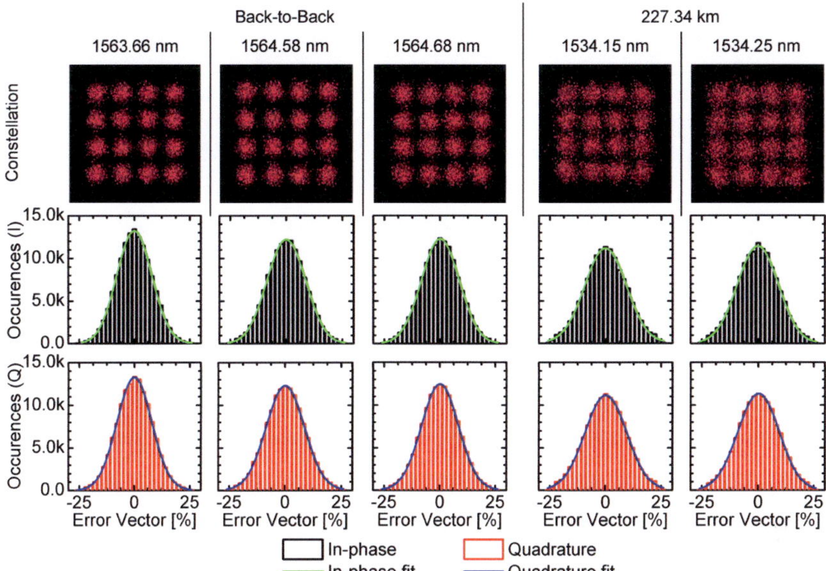

Fig. 6.11: Statistical analysis of selected carriers. We show constellation diagrams of the carriers used for checking the validity of the assumptions for the BER-EVM relationship presented in Eq. (5.17). We performed a statistical analysis of the in-phase (I ——) and quadrature-phase (Q ——) error vectors. The respective Gaussian fits (for I —— and Q ——) indicate a Gaussian probability density function of the added noise as required for the reported BER estimations.

In summary, our results show that a FIR filter with 64 taps suffices for implementing a Nyquist WDM transmission system when using twofold oversampling. However, we expect an additional performance improvement when longer filter lengths are used. The quality of reception could not be compared to 16QAM WDM experiments as linear crosstalk from neighboring carriers limited the EVM. Also, as all carriers in this experiment were derived from a single laser, the experiment should not be compared to experiments using a large number of separate lasers. Here, we were able to increase the data rate by 20 % and the transmission distance by a factor of 4.5 compared to our previous experiment [8] with record-high data rates transmitted on a single laser.

Tab. 6.1: Comparison of estimated and measured BER. We measured BER and EVM for the carriers presented below and calculated an equivalent BER to verify the EVM – BER relationship.

Distance	Wavelength [nm]	EVM [%]	Calculated BER	Measured BER
B2B	1563.66	11.0	1.2×10^{-3}	1.2×10^{-3}
	1564.58	11.9	1.9×10^{-3}	2.1×10^{-3}
	1564.68	12.0	2.0×10^{-3}	2.3×10^{-3}
227 km	1534.15	13.8	5.9×10^{-3}	5.7×10^{-3}
	1534.25	14.8	9.1×10^{-3}	9.4×10^{-3}

6.2.7 Conclusion

In this paper we show that Nyquist WDM is a promising candidate for next generation communication systems. Nyquist WDM improves spectral efficiency and transmission distance when compared to previously investigated all-optical OFDM systems. We demonstrate for the first time 16QAM Nyquist WDM transmission with a symbol rate equal to the carrier spacing. The sinc-pulse shaping is done by real-time digital signal processing. A total aggregate data rate of 32.5 Tbit/s and a net spectral efficiency of 6.4 bit/s/Hz are achieved. As all carriers are generated from a single laser, this is a new data rate record when using a single laser source.

7 Summary and Outlook

Technologies for Single–Laser Tbit/s Communication Systems

For the investigation of Tbit/s communication systems, three key components were developed: An optical multi-format transmitter, an optical frequency comb generator, and an optical FFT.

The Optical Multi–Format Transmitter

An optical multi-format transmitter has been developed. As this transmitter is based on field programmable gate arrays (FPGA) and digital to analog converters (DAC), the transmitted modulation format is defined by the programming of the FPGAs and the format is only limited by the resolution of the DACs. The FPGAs generate the data to be transmitted in real-time and allow operation of the DAC at sampling rates of up to 30 GSa/s. The FPGA features a multi-gigabit pseudo random bit stream generator and a modulation mapper module to generate QAM data in real-time. In this work, the transmitter has been demonstrated to generate OOK, BPSK, QPSK and 16QAM at symbol rates up to 30 GBd in real-time.

To this date, the transmitter has been used for several experiments, of which only a fraction has been published [3, 5-8, 13, 31, 32, 84, 85, 147-149]. The large number of different modulation formats has proven the transmitter to be extremely versatile. In follow up work, the transmitter has been extended and it now supports modulation formats up to 64QAM at 28 GBd [3], OFDM up to 101.5 Gbit/s [4, 31], and advanced pulse shaping methods at up to 150 Gbit/s [5, 6, 32]. In addition to the real-time generation of signals, the transmitter can also be configured to send a preprocessed pattern. Due to the FPGA-based signal processing, even more modulation formats will be implemented in future work.

Optical Comb Generation

A novel scheme has been presented and implemented that is capable of generating 325 optical carriers from a single laser. The scheme generates extremely broad frequency combs with sufficient quality to transmit several tens of Terabit/s. The generated frequency comb covers most of the central communication band. This was made possible by spectral composing of the generated comb to remove the minima in broadened optical spectra generated by self phase modulation [2]. The optical carrier to noise ratio within a resolution bandwidth of 0.1 nm has been found to be between 39.6 dB / 0.1 nm and 25.9 dB / 0.1 nm. In the following, the quality of the carriers has been sufficient to transmit 32.5 Tbit/s over distances up to 227 km.

For future work, implementation of the extension of this scheme as it has been proposed in Fig. 4.3. This extension promises an even larger number of carriers. Second, a more detailed characterization of the generated frequency comb should be performed. As the line-width of the seed MLL is significantly lower than the line-width of standard lasers for optical communication systems, the linewidth of the overall comb could be low as well.

Optical Signal Processing – The Optical Fast Fourier Transform

An optical FFT has been presented that significantly reduced the complexity of the circuit when compared to a direct implementation of the FFT. This FFT has enabled OFDM transmission experiments at data rates of up to 26 Tbit/s. The optical FFT enables real-time processing of signals far beyond the speed of electronics and demonstrates the potential of optical signal processing for future optical communication systems.

Even though the optical FFT has only been applied to OFDM-receivers in this work, it is not limited to this application. A concept for an OFDM transmitter has been presented in Section 5.1.3.1. In addition, one can imagine building flexible add-drop multiplexers similar to what has been done for optical time division multiplexing [150-156]. As the FFT is a very general signal processing block, it could be applied in other areas than optical communication systems. In our experiments, the optical FFT has been implemented using discrete components. To fully exploit the benefits of optical signal processing, integrated optical solution are preferred. Of great interest is silicon photonics, where such an FFT could be integrated with an array of optical modulators, a comb source, and nonlinear optical elements. [129, 157-164].

Terabit/s Transmission System Demonstrations

The components developed during this thesis were pivotal for the implementation of the following transmission experiments:

All–Optical OFDM

An all-optical OFDM transmitter and receiver were designed, implemented and experimentally verified. In these experiments, OFDM signals with data rates of 5.4, 10.8, and 26 Tbit/s were generated in real-time. The 26 Tbit/s signal was transmitted over a distance of 50 km. At the receiver, these signals were demultiplexed using the optical FFT and demodulated in a coherent receiver. Subsequent signal processing was performed offline. For comparison, the signal is also received using narrow filtering with a coherent receiver and a coherent receiver without additional signal processing. The optical FFT clearly outperformed the two other schemes tested in the experiment. In the experiment, an optical FFT with a size of $N = 8$ was used in combination with optical filters. Net spectral efficiencies of 2.7 bit/s/Hz for the 5.4 Tbit/s signal, 4.6 bit/s/Hz for the 10.8 Tbit/s signal and 5.3 bit/s/Hz for the 26.0 Tbit/s signal were achieved. These were the first experiments to show data rates beyond 10 Tbit/s encoded on a single laser [7].

Nyquist WDM

Using Nyquist WDM instead of all-optical OFDM, the data rate and transmission reach were increased further. The experiment was similar to the 26 Tbit/s experiment. However, in contrast to the OFDM approach, the carriers were encoded with sinc-shaped Nyquist pulses that carry 16QAM data at 12.5 GBd. In this experiment, the multi-format transmitter generated sinc-shaped Nyquist pulses in real-time using a finite impulse response filter. The filter length was limited to 32 symbols, which slightly increased the spectral width of the modulated signals. This led to a realistic residual crosstalk that would also be expected in a real implementation of such a system. The signal with 32.5 Tbit/s was transmitted over a distance of 227 km, which is more than 4 times longer than in the 26 Tbit/s experiment. The spectral efficiency was increased to 6.4 bit/s/Hz. This experiment was the first Nyquist WDM experiment with real-time DSP and it demonstrated the highest data rate encoded on a single laser so far [13, 14].

Appendix A: Methods

This methods section contains additional details and derivations for different sections throughout this work. The following topics are covered:

- The derivation of the orthogonality for OFDM and Nyquist pulse shaping first published in [6],

- the optoelectronic modulator types used in this work,

- key optical components and the optical implementation of the coherent receiver,

- quality metrics for optical signals [85, 147, 165],

- some key details on Nyquist pulse shaping [6],

- implementation details of the multi-format transmitter [60], and

- the working principles of the waveshaper [166].

A.1 Derivations of the Orthogonality for OFDM and Nyquist–Pulse Shaping

The following derivations have been published as appendix in [6].

Real–time Nyquist pulse generation beyond 100 Gbit/s and its relation to OFDM

R. Schmogrow, M. Winter, M. Meyer, A. Ludwig, D. Hillerkuss, B. Nebendahl, S. Ben-Ezra, J. Meyer, M. Dreschmann, M. Huebner, J. Becker, C. Koos, W. Freude, and J. Leuthold

Optics Express **20**(1), 317-337 (2012). [6]

At this point we want to describe in mathematical detail the properties of Nyquist signals and illustrate their close relation to OFDM. For a better understanding, Tab. A.1 presents an overview of frequently used symbols contrasting OFDM-specific to Nyquist-specific parameters. As usual, the symbols t and f stand for time and frequency.

Tab. A.1 Commonly used terms for OFDM and Nyquist signal description.

OFDM		Nyquist	
T_s	temporal width of symbol	F_s	spectral width of symbol
F_s	spectral subcarrier spacing	T_s	temporal subcarrier spacing
f_k	spectral subcarrier position	t_k	temporal subcarrier position
$k = 0 \ldots N-1$	index for spectral position of subcarrier (sinusoidal in time)	$k = -\infty \ldots -\infty$	index for temporal position of subcarrier (sinusoidal in frequency)
$i = -\infty \ldots +\infty$	index for temporal position of symbol (rectangular in time)	$i = 0 \ldots N-1$	index for spectral position of symbol (rectangular in frequency)

For general usage within this section we introduce a new set of variables z, Z, m, and Q since the equations can be related either to OFDM or Nyquist signals in frequency or time domain, whichever is of interest. First we define a rectangular window and a sinc-function by

$$\text{rect}\left(\frac{z}{Z}\right) = \begin{cases} 1 & \text{for} |z| < Z/2 \\ 0 & \text{else} \end{cases}, \quad \text{sinc}\left(\frac{z}{Z}\right) = \begin{cases} 1 & z = 0 \\ \dfrac{\sin(\pi z/Z)}{\pi z/Z} & z \neq 0 \end{cases}. \tag{8.1}$$

In the following we summarize the mathematical relations that hold in general:

A.1.1 Orthogonality relations

$$\frac{1}{Z}\int_{-Z/2}^{+Z/2} e^{j2\pi mz/Z}\, e^{-j2\pi m'z/Z}\, \mathrm{d}z = \delta_{mm'} \qquad \text{for} \qquad m, m' \in \mathbb{Z} \tag{8.2}$$

$$\frac{1}{Z}\int_{-\infty}^{+\infty} \text{sinc}\left(\frac{z}{Z}-m\right)\text{sinc}\left(\frac{z}{Z}-m'\right)\mathrm{d}z = \delta_{mm'} \qquad \text{for} \qquad m, m' \in \mathbb{Z} \tag{8.3}$$

A.1.2 Series expansions

We expand functions $\varphi(z)$ in a series of orthogonal complex harmonics with the help of the orthogonality relation Eq. (8.2), and functions $\psi(z)$ in a series of orthogonal sinc-functions observing the orthogonality relation Eq. (8.3),

$$\varphi(z) = \sum_{m=-\infty}^{+\infty} \varphi_m\, e^{j2\pi mz/Z}, \tag{8.4}$$

$$\varphi_m = \frac{1}{Z}\int_{-Z/2}^{+Z/2} \varphi(z)e^{-j2\pi mz/Z}\, \mathrm{d}z \tag{8.5}$$

$$\psi(z) = \sum_{m=-\infty}^{+\infty} \psi_m \, \mathrm{sinc}\left(\frac{z}{Z} - m\right) \tag{8.6}$$

$$\psi_m = \frac{1}{Z} \int_{-\infty}^{+\infty} \psi(z) \, \mathrm{sinc}\left(\frac{z}{Z} - m\right) dz \tag{8.7}$$

A.1.3 Power relations

$$\frac{1}{Z} \int_{-Z/2}^{+Z/2} |\varphi(z)|^2 \, dz = \sum_{m=-\infty}^{+\infty} |\varphi_m|^2, \tag{8.8}$$

$$\frac{1}{Z} \int_{-\infty}^{+\infty} |\psi(z)|^2 \, dz = \sum_{m=-\infty}^{+\infty} |\psi_m|^2. \tag{8.9}$$

A.1.4 Peak power of a sum of sinc–functions

Nyquist signals and OFDM spectra are both described by a sum $s(z)$ of equidistantly shifted sinc-functions, Eq. (8.6). We are interested in a worst-case estimation of the maximum power $|s_{max}|^2$. To this end we assume a constant height of all sinc-functions by choosing coefficients $|\psi_m| = 1$ with equal magnitude. The signs of the coefficients ψ_m are then selected such that a maximum $s_{max}(z_{max})$ is found at some position z_{max}. We start by expanding the special function $s^{(1)}(z) = 1$ in a series of sinc-functions, Eq. (8.6). The expansion coefficients ψ_m are calculated to be $\psi_m = 1 \, \forall m$ by evaluating Eq. (8.7) and observing that ([167], Vol. 1, p. 454, formula 3.721 1.)

$$\frac{1}{Z} \int_{-\infty}^{+\infty} \mathrm{sinc}\left(\frac{z}{Z} - m\right) dz = 1. \tag{8.10}$$

From Eq. (8.6) it follows that

$$s^{(1)}(z) = \sum_{m=-\infty}^{+\infty} \mathrm{sinc}\left(\frac{z}{Z} - m\right) = 1. \tag{8.11}$$

Eq. (8.11) shows that performing a summation of equally spaced sinc-functions with identical weight leads to a value of 1 at any position z. This value can be exceeded by choosing the expansion coefficients ψ_m appropriately. For this it should be noted that the sinc-function flips sign between adjacent intervals bounded by zeros. The maximum value of the sum $s(z)$ is obtained when all sinc-functions have the same sign in the z-interval under consideration. This is true for

$$\begin{aligned} s(z) &= \sum_{m=-\infty}^{+\infty} \psi_m \, \mathrm{sinc}\left(\frac{z}{Z} - m\right) \\ &= \sum_{m=0}^{+\infty} (-1)^m \, \mathrm{sinc}\left(\frac{z}{Z} - m\right) + \sum_{m=-\infty}^{-1} (-1)^{m+1} \, \mathrm{sinc}\left(\frac{z}{Z} - m\right), \end{aligned} \tag{8.12}$$

where the coefficients ψ_m have been chosen such that pairs of sinc-functions $m = (0, +1); (-1, +2); (-2, +3); \ldots$ have all a positive sign in the interval $0 < z < Z$. The resulting function $s(z)$ is monotonic in $0 < z < Z$ and symmetrical with respect to $z = Z/2$, so that the superposition of each pair has its maximum at this point, as will be explained in the following.

Consider a function $f(u)$ which is monotonic in an interval $-U < u < +U$ ($U > 0$). In this interval the sum $s_f(u) = f(u) + f(-u)$ has an extremum if $s_f'(u) = f'(u) - f'(-u) = 0$, i. e., for $u = 0$. This result as applied to Eq. (8.12) means that the maximum is found at the symmetry point $z_{max} = Z/2$ of the sum $s(z)$,

$$s_{max}\left(\tfrac{1}{2}Z\right) = \sum_{m=-\infty}^{+\infty} \left|\operatorname{sinc}\left(\tfrac{1}{2} - m\right)\right| \tag{8.13}$$

Note that the sum does not converge. However, Eq. (8.13) also applies to a finite sum with a maximum of Q sinc-functions, from which the maximum power $s_{Q,max}^2(Z/2)$ can be computed,

$$s_{Q,max}\left(\tfrac{1}{2}Z\right) = \sum_{m=-Q/2+1}^{Q/2} \left|\operatorname{sinc}\left(\tfrac{1}{2} - m\right)\right| \tag{8.14}$$

A.1.5 Average power of an oversampled sinc–function

For deriving the average power of a sum of oversampled shifted sinc-functions $\operatorname{sinc}(qz/Z{-}m)$ (oversampling factor q), we expand $\psi(z) = \operatorname{sinc}(z/Z)$ Eq. (8.6), but this time in terms of oversampled sinc-functions $\operatorname{sinc}(qz/Z{-}m)$. We find the expansion coefficients $\psi_m = \operatorname{sinc}(m/q)$ according to Eq. (8.7) and write

$$\psi(z) = \operatorname{sinc}\left(\frac{z}{Z}\right) = \sum_{m=-\infty}^{+\infty} \operatorname{sinc}\left(\frac{m}{q}\right) \operatorname{sinc}\left(q\frac{z}{Z} - m\right). \tag{8.15}$$

By substituting $\psi(z)$ in the power relation Eq. (8.9) and by applying the orthogonality relation Eq. (8.3) we find the average power

$$\overline{P} = \frac{1}{Z}\int_{-\infty}^{+\infty} |\psi(z)|^2 \, dz = \frac{1}{q}\sum_{m=-\infty}^{+\infty} \operatorname{sinc}^2\left(\frac{m}{q}\right) = 1. \tag{8.16}$$

In real life, oversampling the base functions by a factor q (preferably $q = 2$) is needed to simplify the filtering of a Nyquist channel. The $\operatorname{sinc}(z/Z)$-function is then represented not by a number of Q base functions as in Eq. (8.14), but by $q\,Q$ base-functions, and again orthogonality is lost in the strict sense. Nevertheless we approximate Eq. (8.15) by

$$\psi(z) = \operatorname{sinc}\left(\frac{z}{Z}\right) \approx \sum_{m=-qQ/2+1}^{+qQ/2} \operatorname{sinc}\left(\frac{m}{q}\right) \operatorname{sinc}\left(q\frac{z}{Z} - m\right). \tag{8.17}$$

If $q\,Q$ is large enough, the average power should be still close to 1,

$$P^{(q,Q)} = \frac{1}{q}\sum_{m=-qQ/2+1}^{+qQ/2} \operatorname{sinc}^2\left(\frac{m}{q}\right) \approx 1. \tag{8.18}$$

In reality we not only have a finite number $q\,Q$ of base functions, but the so far assumed equal modulus for all expansion coefficients must be modified if QAM modulated signals come into play. In this case, the approximated average power Eq. (8.18) needs to be divided by a format dependent factor k^2 [85, 147], which relates the maximum power of the constellation points to the mean power for all constellation points. Therefore we write approximately

$$\overline{P} \approx P^{(q,Q,k)} = \frac{P^{(q,Q)}}{k^2} = \frac{1}{k^2 q} \sum_{m=-qQ/2+1}^{qQ/2} \operatorname{sinc}^2\left(\frac{m}{q}\right) \approx 1. \tag{8.19}$$

A.1.6   PAPR for a Nyquist signal

The average power in Eq. (8.19) serves as reference for the PAPR whereas the maximum power is determined by Eq. (8.14). We obtain

$$\text{PAPR}_{\text{Nyquist}} = \frac{s_{Q,\max}^2\left(\frac{1}{2}Z\right)}{\overline{P}} = k^2 \frac{\left[\sum_{m=-Q/2+1}^{Q/2} \left|\operatorname{sinc}\left(\frac{1}{2}-m\right)\right|\right]^2}{\frac{1}{q}\sum_{m=-qQ/2+1}^{+qQ/2} \operatorname{sinc}^2\left(\frac{m}{q}\right)}. \tag{8.20}$$

This equation corresponds to Eq. (8.48) in the main body of this paper.

A.1.7   Peak power of an OFDM symbol

An OFDM symbol with Q sinusoidal carriers constant within a window of width Z be given by

$$s(z) = \sum_{m=1}^{Q} \sqrt{2}\cos\left(2\pi m \frac{z}{Z} + \alpha_m\right). \tag{8.21}$$

If the phases α_m of the Q carriers are chosen accordingly and all symbols have maximum values, then all amplitudes add up leading to:

$$s_{Q,\max} = \sqrt{2}\,Q. \tag{8.22}$$

A.1.8   Average power of an OFDM symbol

The average power can be determined with the power relation Eq. (8.8),

$$P^{(Q)} = \frac{1}{Z}\int_{-Z/2}^{+Z/2} |s(z)|^2 \, dz = \sum_{m=1}^{Q} \frac{2}{Z}\int_{-Z/2}^{+Z/2} \cos^2\left(2\pi m \frac{z}{Z} + \alpha_m\right) dz$$
$$= \sum_{m=1}^{Q} \frac{2}{Z}\frac{1}{2}Z = \sum_{m=1}^{Q} 1 = Q. \tag{8.23}$$

Strictly speaking, orthogonality is lost if the OFDM spectrum is truncated as is always the case in reality. Nevertheless, Eq. (8.23) represents a good approximation for the average power of an OFDM signal comprising a sufficient number of Q subcarriers. Similar to the arguments leading to Eq. (8.19), the average power in a symbol needs to be divided by a format dependent factor k^2 [85, 147] such that the average power in a symbol is

$$\overline{P}^{(Q,k)} = Q / k^2 \tag{8.24}$$

A.1.9 PAPR of an OFDM symbol

The PAPR follows by relating Eq. (8.22) to Eq. (8.24). We find

$$\mathrm{PAPR}_{\mathrm{OFDM},k=1} = \frac{s_{Q,\mathrm{max}}^2}{\overline{P}^{(Q,k)}} = 2k^2 Q \tag{8.25}$$

For Eq. (8.25) the same number of elementary functions was adopted as for Eq. (8.20).

A.1.10 Spectrum of a Nyquist signal

The spectrum $Y_{\mathrm{FIR}}^{(0)}(f,R)$ of a Nyquist signal having a finite extent in time results from convolving a rectangular spectrum $Y^{(0)}(f)$ of Eq. (2.15) (representing the spectrum symmetrical to $f = 0$ of an infinitely extended baseband Nyquist sinc-impulse) with a sinc-shaped spectrum $W(f,R)$ (representing the spectrum of a rectangular time window $w(t) = \mathrm{rect}[t / (RT_s / q)]$ which depends on the number of filter taps R and the oversampling factor q),

$$Y_{\mathrm{FIR}}^{(0)}(f,R) = Y^{(0)}(f) * W(f,R) = T_s \, \mathrm{rect}\!\left(\frac{f}{F_s}\right) * \frac{R}{q} T_s \, \mathrm{sinc}\!\left(R\frac{f}{qF_s}\right). \tag{8.26}$$

On evaluation we find in terms of the sine integral [168] $\mathrm{Si}(z) = \int_0^z (\sin v / v)\, \mathrm{d}v$

$$Y_{\mathrm{FIR}}^{(0)}(f,R) = \frac{T_s}{\pi}\left[\mathrm{Si}\!\left(\pi R \frac{f + F_s/2}{qF_s}\right) - \mathrm{Si}\!\left(\pi R \frac{f - F_s/2}{qF_s}\right)\right]. \tag{8.27}$$

A.2 Optoelectronic Modulators

A.2.1 The Mach–Zehnder Modulator

The Mach-Zehnder modulator (MZM) [169] is the most commonly used modulator in optical communication systems. It allows for amplitude and phase modulation of the optical field. Such a modulator is based on a Mach-Zehnder interferometer [170, 171], in which the phase of one or both arms is modulated (see Fig. A.1a).

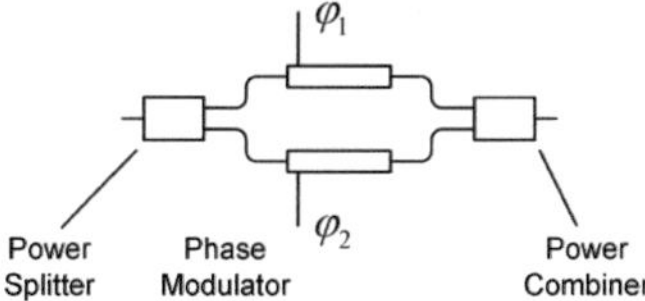

Fig. A.1. In the Mach-Zehnder modulator the incoming carrier or signal is split into two arms, the two arms are modulated in phase and recombined afterwards to interfere, which enables phase and amplitude modulation.

The light is split, the two paths are modulated in phase (φ_1, φ_2) and the signal is afterwards recombined. The interference of these two paths then defines the resulting amplitude and phase. The output can then be calculated according to eq. (8.28).

$$E_{\text{out}} = \sqrt{\frac{1}{2}} \left(\sqrt{\frac{1}{2}} \left(e^{j(\varphi_1)} + e^{j(\varphi_2)} \right) E_{\text{in}} \right) \tag{8.28}$$

From eq. (8.28), one can now calculate the transfer function:

$$H = \frac{E_{\text{out}}}{E_{\text{in}}}$$

$$= \frac{1}{2} \left(e^{j(\varphi_1)} + e^{j(\varphi_2)} \right) \tag{8.29}$$

$$= \underbrace{e^{j\frac{\varphi_1+\varphi_2}{2}}}_{\text{Phase Modulation}} \times \underbrace{\cos\left(\frac{\varphi_1-\varphi_2}{2}\right)}_{\text{Amplitude Modulation}}$$

From this transfer function, one can see that the output amplitude depends on the phase difference of the two arms, while the phase is the average phase of the two arms. With this, all phase and amplitude states can be achieved. One can now define two specific operating modes of the modulator. In push-push mode, the two phases are equal $\left(\varphi_1 = \varphi_2\right)$ resulting in a pure phase modulation, while the two phases are inverted $\left(\varphi_1 = -\varphi_2\right)$ in push-pull mode, leading to a pure amplitude modulation.

A.2.1.1 Transfer Function of the Mach–Zehnder–Modulator

When operating the Mach-Zehnder Modulator in push-pull configuration, the modulator transfer function (8.29) can be simplified. The phase modulation term vanishes and the equation may be rewritten:

$$H = \underbrace{e^{j\frac{\varphi_1+\varphi_2}{2}}}_{\text{Phase Modulation}} \times \underbrace{\cos\left(\frac{\varphi_1-\varphi_2}{2}\right)}_{\text{Amplitude Modulation}} \Bigg\| \quad \varphi = \varphi_1 = -\varphi_2 \tag{8.30}$$

$$= \cos\left(\varphi\right)$$

This transfer functions for field and intensity are plotted in Fig. A.2 to illustrate the dependence of the optical signal at output of the Mach-Zehnder modulator to the introduced differential phase shift.

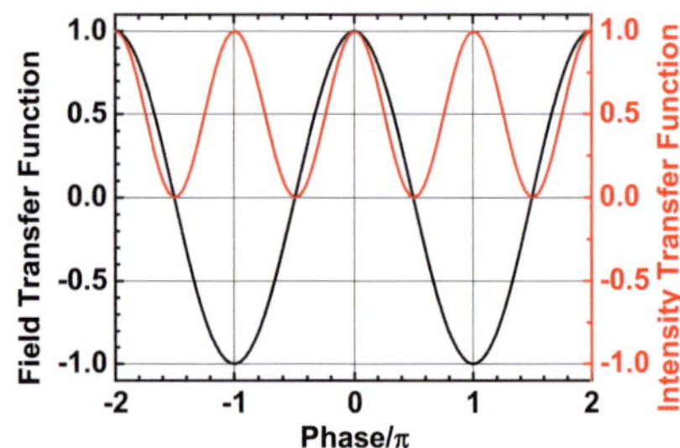

Fig. A.2: Normalized field and intensity transfer functions of a Mach-Zehnder modulator. The field transfer function (———) is of the form cos(φ), while the intensity transfer function (———) is of the form cos^2(φ). Where the field transfer function is negative, a phase shift of 180° is introduced. The output amplitude is given by the absolute value of the field transfer function.

A.2.2   The optical IQ modulator

Even though one can achieve any phase and amplitude with a dual drive Mach-Zehnder modulator, an optical IQ modulator is often chosen instead [172].

A.2.2.1 Design and Functionality

An optical I/Q-Modulator consists of two Mach Zehnder Modulators (MZM – see A.2.1) and a 90° phase shifting element. Both MZM are aligned parallel in an in-phase (I) and a quadrature phase (Q) arm as depicted in Fig. A.3. The 90° phase shift is introduced in the Q arm of the modulator.

The optical carrier at the input of the modulator is split into two arms. Each arm contains a separate Mach-Zehnder modulator (MZM 1 and MZM 2). These modulators are operated as amplitude modulators in push-pull configuration ($\varphi_{I,1} = -\varphi_{I,2}$ and $\varphi_{Q,1} = -\varphi_{Q,2}$). The first arm is not shifted in phase, such it is called in-phase (I). An additional 90° phase shift in the second arm is introduced, making it the quadrature component (Q). Through this phase shift, the two outputs that can be represented as cosine (I) and sine waves (Q) are orthogonal with respect to each other (8.31).

$$\int_0^T \cos\left(2\pi f \cdot t\right) \cdot \left(-\sin\left(2\pi f \cdot t\right)\right) dt = 0 \tag{8.31}$$

by properly adjusting the amplitudes I and Q, every position of the complex plane (Fig. A.3 – right hand side) can be reached. More details on that follow in the upcoming section (A.2.1.1).

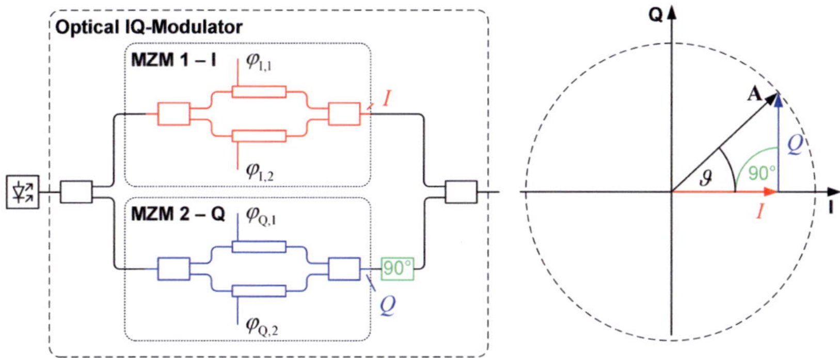

Fig. A.3: Schematics and principle of operation of an optical IQ-modulator. In an optical IQ-modulator, the optical carrier is split into two arms. Each arm contains a Mach-Zehnder modulator (MZM 1 and MZM 2). Both modulators are operated in push-pull configuration for pure amplitude modulation. A 90° phase shift (90°) is introduced at the output of MZM 2 before the two MZM outputs are recombined. The outputs of the two MZMs are phasors in the complex plane and can be plotted in the IQ-diagram or constellation diagram (see right hand side). The phasors are defined by the two components I for MZM 1 and Q for MZM 2 and the relative phase shift of 90°. As I and Q can have positive and negative values, the output phase φ_{out} can be set freely and the output amplitude $|\mathbf{A}|$ is only limited by the input power to the modulator.

A.2.2.2 Transfer function of the IQ modulator

In this expression the index x stands for I or Q respectively.

$$\varphi_{x1} = -\varphi_{x2} = \varphi_x \rightarrow \underline{T}_s = j \cdot \cos(\varphi_x) \tag{8.32}$$

Taking into account the 90° phase shift to satisfy the orthogonality condition described in (8.31), the analytic transfer function of the whole modulator for a complex optical field can be derived.

$$\underline{T}_{total} = \frac{1}{2} \cdot \begin{bmatrix} 1 & j \end{bmatrix} \cdot \begin{bmatrix} j\cos(\varphi_I) & 0 \\ 0 & j \cdot j\cos(\varphi_Q) \end{bmatrix} \cdot \begin{bmatrix} j \\ 1 \end{bmatrix} = -\frac{1}{2} \cdot \left(\cos(\varphi_I) + j\cos(\varphi_Q) \right) \tag{8.33}$$

From (8.33) it can again be seen that any position in complex plane may be reached using the setup described in A.2.2.1.

For a real optical carrier, the real electrical field at the output of the modulator is described by (8.34).

$$E_{out}(t) = \frac{1}{\sqrt{2}} \cdot \left[\cos(\varphi_I) \cdot \cos(2\pi f_c \cdot t + \Phi) - \cos(\varphi_Q) \cdot \sin(2\pi f_c \cdot t + \Phi) \right] \tag{8.34}$$

Here parameter Φ includes an initial phase of the laser and an overall phase delay introduced in the modulator. This common phase term does not violate (8.31) since both carrier waves are always orthogonal with respect to each other as long as the 90° phase difference between is maintained (8.35).

$$\int_0^T \cos(\omega \cdot t + \Phi) \cdot \sin(\omega \cdot t + \Phi)\, dt = 0 \qquad (8.35)$$

In that way it is always possible to separate the data encoded on in-phase and quadrature component at the receiver.

A.2.3 The Electro–Absorption Modulator

Electro-absorption modulators (EAM) are of great interest for applications, where the light has to be modulated independently of the polarization. After transmission, the polarization of the incident light is often unknown and also changes constantly due to drift. In contrast to Mach-Zehnder modulators based on Lithium Niobate ($LiNbO_3$), EAMs are only slightly polarization dependent. So they are the natural choice for gating elements in optical receivers.

EAMs are based on a reverse biased laser section with an additional modulation of the reverse voltage. The electrical field changes the absorption of the EAM through the quantum-confined Stark effect and the Franz-Keldysh effect. The dominant effect in such quantum well structures is the quantum-confined Stark effect [173], which describes the change in band-edge absorption due to the deformation of the quantum wells induced by the electrical field. The Franz-Keldysh effect [174] is also present in such modulators, even though it is significantly smaller than the quantum confined stark effect [173]. The Franz-Keldysh effect also describes a broadening of the band-edge absorption due to photon assisted tunneling of valence electrons to the conduction band through virtual states within the bandgap [174]. This is possible due to the deformation of the band structure when an external electrical field is applied.

Fig. A.4 shows the transfer function of the electro absorption modulator used in our experiments. This modulator offers a high extinction for all wavelengths and works well for both polarizations. The transfer function of this modulator strongly depends on the wavelength of the modulated light, which is expected as the effects that are exploited depend on the photon energy in relation to the band gap of the semiconductor.

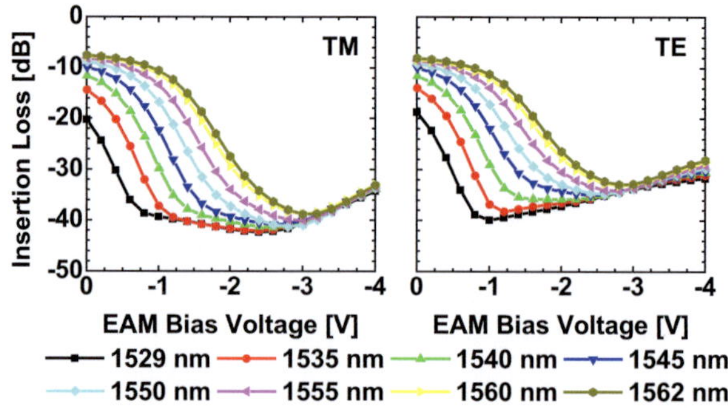

Fig. A.4: Transfer function on the CIP-40G-PS-EAM electro absorption modulator for TM (left) and TE polarization (right) [151].

A.3 The Optical Coherent Receiver

Fig. A.5 shows the implementation of a coherent receiver for optical communication systems.

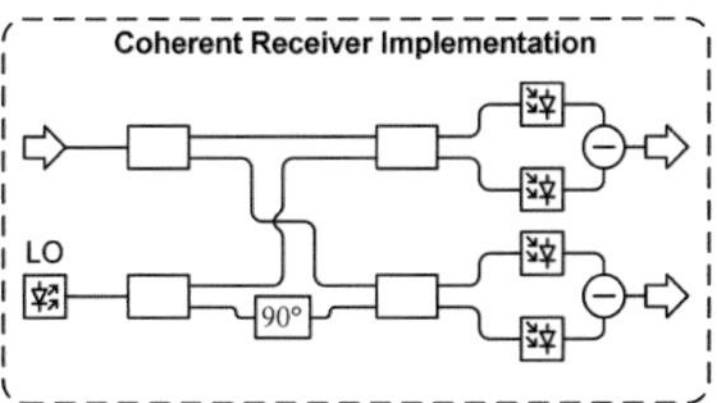

Fig. A.5: Implementation of a coherent receiver for optical communication systems. The incident signal is mixed with the local oscillator and a copy of the local oscillator that has been shifted by 90°.

A.4 Quality Metrics for Optical Signals

The following section has been published as [165], which includes work first presented in [85, 147]. It has been modified to include an additional figure (Fig. A.8) from [85]. In Addition, a misprint [147] was corrected in equation (4) of reference [165], here equation (8.40) where a factor of $12 / M - 1$ was missing.

Quality Metrics for Optical Signals: Eye Diagram, Q–factor, OSNR, EVM and BER

W. Freude, R. Schmogrow, B. Nebendahl, M. Winter, A. Josten, D. Hillerkuss, S. Koenig, J. Meyer, M. Dreschmann, M. Huebner, C. Koos, J. Becker, and J. Leuthold

in International Conference on Transparent Optical Networks (ICTON) (2012) [165].

Abstract: Measuring the quality of optical signals is one of the most important tasks in optical communications. A variety of metrics are available, namely the general shape of the eye diagram, the optical signal-to-noise power ratio (OSNR), the Q-factor as a measure of the eye opening, the error vector magnitude (EVM) that is especially suited for quadrature amplitude modulation (QAM) formats, and the bit error ratio (BER). While the BER is the most conclusive quality determinant, it is sometimes difficult to quantify, especially for simulations and off-line processing. We compare various metrics analytically, by simulation, and through experiments. We further discuss BER estimates derived from OSNR, Q-factor and EVM data and compare them to measurements employing six modulation formats at symbol rates of 20 GBd and 25 GBd, which were generated by a software-defined transmitter. We conclude that for optical channels with additive Gaussian noise the EVM metric is a reliable quality measure. For nondata-aided reception, BER below 0.01 can be estimated from measured EVM.

A.4.1   Introduction

The ultimate quality measure in optical communication links is the bit error ratio (BER, bit error probability). However, a direct experimental BER determination can be done only as long as the link is out of service (if not a hard-decision forward error correction (FEC) is able to report errors), because a known data sequence has to be transmitted, and it consumes a significant amount of time if the BER is small. If the estimated $BER = 10^{-r}$ (which itself is a random number) should lie with probability $\alpha = 99\,\%$ in a confidence interval $\varepsilon = (BER_{max} - BER) / BER = 100\,\%$, a number of 13 bit errors has to be found [175]. For a data rate BR, a single BER measurement takes the time $\tau_{BER} = 13 \times 10^{r}\,bit\,/\,BR$. For $r = 4$ ($r = 12$) and $BR = 10$ Gbit/s we find $\tau_{BER} = 13$ µs ($\tau_{BER} = 1\,300$ s). For simulations and off-line signal processing the effective data rate is smaller by many orders of magnitude, so counting errors becomes impractical [175-177]. It is therefore tempting to estimate the BER based on a few measured moments of the probability density function (pdf), which describes the noise at the decision circuit of the receiver. — In the following, we discuss various moment-based metrics and compare simulations with experiments.

A.4.2   Eye Diagram and Q–Factor for On–Off Keying Signals and Direct Reception

A fairly general receiver schematic for on-off-keying (OOK) signals is depicted in Fig. A.6(top). It consists of an optical pre-amplifier (OA) and a filter with a bandwidth B_O having an output electric field strength E. A photodetector transforms the time average $<|E|^2>$ over a few optical periods into a proportional photocurrent, which is converted into a voltage by an electronic amplifier (EA). After a filter with bandwidth B_E the output voltage u has to be associated with a one or a zero.

First, we look at a receiver without optical pre-amplifier. For small received signal powers $<|E|^2>$ the noise is mostly due to the EA. This situation is depicted in form of an eye diagram of the voltage u as a function of time t. The eye shows very little intersymbol interference (ISI). Detailed methods to analyze the eye were reported recently [178]. We operate at the optimum sampling time, Fig. A.6(bottom). Significant noise will be recorded. According to the measured histograms the fluctuating voltage is distributed symmetrically around the average values u_1 for a one and u_0 for a zero. The probability density functions (pdf) of these noise voltages are depicted in Fig. A.7 and are denoted with $w_1(u)$ and $w_0(u)$, respectively.

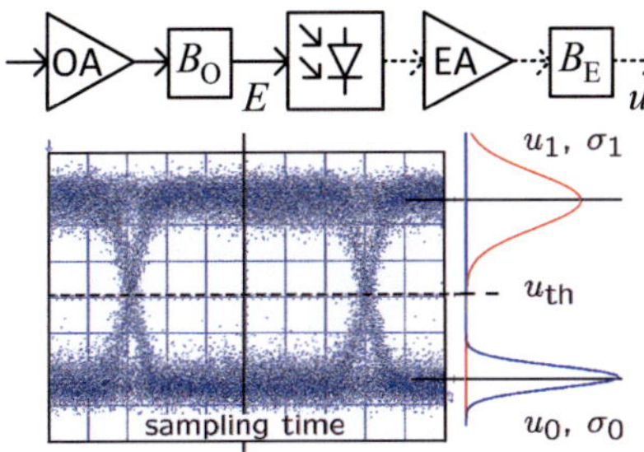

Fig. A.6: Direct OOK receiver and eye diagram. **(top)** Optical pre-amplifier (OA), filter (bandwidth B_O) and electrical field E. Photodiode with electrical amplifier (EA), filter (bandwidth B_E) and output voltage u. **(bottom)** Eye diagram of voltage u with schematic noise histograms (pdf) at mean one level u_1 (standard deviation σ_1, red) and mean zero level u_0 (σ_0, blue).

The probability of transmitting a one or a zero is $p(1\mathrm{t})$ or $p(0\mathrm{t})$, and the BER corresponds to the sum of the appropriately weighed hatched areas in Fig. A.7. The optimum decision threshold $u_\mathrm{th,op}$ for a minimum $\mathrm{BER_{op}}$ is found by calculating $\mathrm{d\,BER}/\mathrm{d}u_\mathrm{th} = 0$,

$$\mathrm{BER} = p(1\mathrm{t})\int_{-\infty}^{u_\mathrm{th}} w_1(u)\,\mathrm{d}u + p(0\mathrm{t})\int_{u_\mathrm{th}}^{+\infty} w_0(u)\,\mathrm{d}u,$$

$$p(1\mathrm{t})w_1(u_\mathrm{th,op}) = p(0\mathrm{t})w_0(u_\mathrm{th,op}).$$

(8.36)

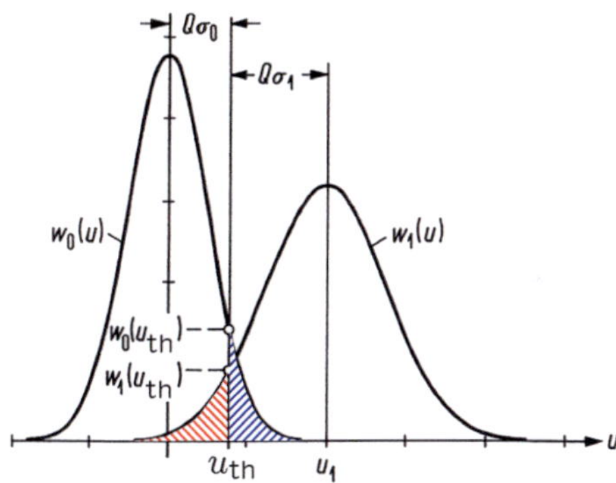

Fig. A.7: Probability density functions $w_{0,1}(u)$ of output noise voltage at mean one and zero levels $u_{1,2}$ (variances $\sigma_{1,2}$) with decider threshold voltage u_th. The hatched blue area ($u > u_\mathrm{th}$) represents the probability $p(1\mathrm{r}|0\mathrm{t})$ to actually receive a one if a zero was transmitted, while the hatched red area ($u < u_\mathrm{th}$) is the probability $p(0\mathrm{r}|1\mathrm{t})$ to receive a zero if a one was transmitted

If ones and zeros are transmitted with equal probability, $p(1\mathrm{t}) = p(0\mathrm{t})$, and the pdf are unimodal, the intersection of $w_1(u)$ and $w_0(u)$ marks the optimum threshold. For an estimate of the BER, both pdf have to be known, and if the BER is small, then the pdf must be accurately known especially in the wings. If the EA noise dominates, the statistics can be well approximated by two Gaussian pdf with mean $u_{1,0}$ and standard deviation $\sigma_{1,0}$, namely $w_{1,0}(u) = 1/(\sqrt{2\pi}\,\sigma_{1,0})\exp[(u - u_{1,0})^2/(2\sigma_{1,0}^2)]$. This approximation is well justified by the central limit theorem [179], which states that the pdf of a sum of independent random variables (represented by the output of filter B_E) tends towards a Gaussian. The complementary error function $\mathrm{erfc}(z) = (2/\sqrt{\pi})\int_z^\infty \exp(-t^2)\,dt$ allows to formulate a simple analytic expression for the optimum $\mathrm{BER_{op}}$, if the ratio of the occurrence probabilities for ones and zeros equals the ratio of the corresponding standard deviations,

$$\mathrm{BER}_{\mathrm{op}} = \frac{1}{2}\,\mathrm{erfc}\!\left(\frac{Q_F}{\sqrt{2}}\right),$$

$$Q_F = \frac{u_1 - u_0}{\sigma_1 + \sigma_0} \approx \sqrt{\gamma},$$

$$u_{\mathrm{th}} = \frac{u_1\sigma_0 + u_0\sigma_1}{\sigma_0 + \sigma_1},$$

$$\frac{w_1(u_{\mathrm{th,op}})}{\sigma_0} = \frac{w_0(u_{\mathrm{th,op}})}{\sigma_1} \quad \text{for} \quad \frac{p(1t)}{p(0t)} = \frac{\sigma_1}{\sigma_0}.$$

$$(8.37)$$

With a sampling oscilloscope we record the random voltage u at the given (optimum) sampling time. From these data the four moments $u_{1,0}$ and $\sigma_{1,0}$ are computed, so that $u_{\mathrm{th,op}}$ and $\mathrm{BER_{op}}$ can be estimated from Eq. (8.37). It is worth noting that $u_{1,0}$ and $\sigma_{1,0}$, being calculated from a statistical sample only, are themselves random variables, and this property is inherited by the BER estimate. All measured quantities involved are either first moments $u_{1,0}$ or central second moments $\sigma_{1,0}^2$ of the random voltage sample u in Fig. A.6.

The so-called Q-factor Q_F in Eq. (8.37), a quality metric that can be defined by moments up to the order two (even if the actual pdf are not known), is related to the electrical signal-to-noise power ratio $\gamma = P_{S,a}/P_N$ under the following assumptions: The received pulses have no ISI, ones and zeros are uniformly distributed, a zero has the mean value $u_0 = 0$, the relation $1 < \sigma_1^2/\sigma_0^2 \leq 2$ holds, the average noise power $P_N = (\sigma_1^2 + \sigma_0^2)/2$ is limited to $\sigma_1^2 < P_N \leq 1.5\sigma_1^2$. The signal power of *one* typical voltage impulse shape with maximum $h(0) = 1$ (e. g., a raised cosine) is $P_S^{(1)} = \langle u_1^2 h^2(t)\rangle \approx u_1^2/2$, which leads to an *average* signal power $P_{S,a} = P_S^{(1)}/2 \approx u_1^2/4$. We then find $\gamma \approx (\frac{1}{3}...\frac{1}{4})u_1^2/\sigma_1^2$ and $Q^2 \approx (\frac{1}{2.91}...\frac{1}{4})u_1^2/\sigma_1^2$ resulting in $\gamma \approx (0.97...1)Q^2 \approx Q^2$ [180]. Therefore a (not very accurate) BER estimate can be derived from a measured electrical signal-to-noise power ratio $\gamma \approx Q^2$.

A.4.3 Optical Signal–to–Noise Power Ratio (OSNR)

For increasing the receiver sensitivity, we add an optical pre-amplifier (OA, power gain $G \gg 1$, spontaneous emission factor n_{sp}, noise figure $F \approx 2n_{\mathrm{sp}} \geq 2$), Fig. A.6(top). In this case the noise of the EA may be usually neglected. The optical signal output power GP_O (photon energy $hf_O = hc/\lambda_O$) and the amplified spontaneous emission (ASE, polarized noise power $P_{\mathrm{ASE}}^{\mathrm{x,pol}} = \frac{1}{2}GFhf_O B_{\mathrm{x}}$ in bandwidth B_{x}) are measured with an optical spectrum analyzer (OSA) and lead to various definitions for signal-to-noise power ratios,

$$\mathrm{OSNR}_{\mathrm{ref}} = \frac{GP_{\mathrm{O}}}{2P_{\mathrm{ASE}}^{\mathrm{ref,\,pol}}} = \frac{P_{\mathrm{O}}}{Fhf_{\mathrm{O}}B_{\mathrm{ref}}},$$

$$\mathrm{OSNR} = \frac{GP_{\mathrm{O}}}{P_{\mathrm{ASE}}^{\mathrm{O,\,pol}}} = \frac{P_{\mathrm{O}}}{\frac{1}{2}Fhf_{\mathrm{O}}B_{\mathrm{O}}} = \frac{2B_{\mathrm{ref}}}{B_{\mathrm{O}}}\mathrm{OSNR}_{\mathrm{ref}}, \qquad (8.38)$$

$$\gamma_{\mathrm{OA}} = \frac{P_{\mathrm{O}}}{Fhf_{\mathrm{O}}B_{\mathrm{O}}} \leq \frac{\mathrm{OSNR}}{2}.$$

The usual definition $\mathrm{OSNR}_{\mathrm{ref}}$ relates the signal power to the unpolarized ASE power in a fixed bandwidth $B_{\mathrm{ref}} = 0.1\,\mathrm{nm} \times c / \lambda_{\mathrm{O}}^2$ (vacuum speed of light c), where $B_{\mathrm{ref}} = 12.5\,\mathrm{GHz}$ at $\lambda_{\mathrm{O}} = 1.55\,\mu\mathrm{m}$. The definition OSNR [64] bears more physical relevance, because it refers the signal power to the polarized ASE power in the actual optical receiver bandwidth B_{O}. For coherent reception, the fields for signal and Gaussian ASE noise are just "linearly" frequency down-converted by the photodiode, but remain otherwise essentially unchanged. If ASE noise dominates over any noise from local laser oscillator and EA, then OSNR and the electrical signal-to-noise ratio are same. However, OSNR from Eq. (8.38) and γ from Eq. (8.37) relate to different scenarios (coherent vs. direct reception, OA vs. EA noise) and cannot be compared in general.

For direct OOK reception in a bandwidth $B_{\mathrm{E}} = B_{\mathrm{O}} / 2$ (correlated upper and lower optical signal sidebands are transmitted in B_{O}), the shot-noise limited electrical signal-to-noise ratio γ_{OA}, Eq. (8.38), is half the measured OSNR due to mixing of ASE noise and signal in the photodiode. Compared to the signal-to-noise ratio at the OA input, γ_{OA} is smaller by a factor $F \geq 2$. If γ_{OA} is measured, one could try estimating the BER with Eq. (8.37).

However, this procedure is problematic. While the ASE contribution to the optical electric field can be well represented by additive white Gaussian noise (AWGN), the electrical noise voltage u after the square-law photodetector is definitely not Gaussian. If a one with mean u_1 is sampled and if σ_1 is not too large, then $w_1(u)$ can be still approximated by a Gaussian pdf, but a sampled zero $u_0 = 0$ results in an exponential pdf $w_0(u)$ [181]. The actual situation is even more complicated when the filtering of $u(t)$ or EA noise are taken into account [52, 182, 183]. Because the photo current is in proportion to the fluctuating optical power $<|E|^2>$, the filter B_{E} sums in effect the squares of the Gaussian distributed electric field samples E, and this leads to a chi-square pdf for u. A similar *caveat* holds if the optical noise is not Gaussian, e. g., if phase noise impairs phase-modulated signals [52]. Endeavours to measure more than two moments, and to approximate the true pdf by a series expansion of Gauss-Hermite or Gauss-Laguerre functions [184, 185] do not truly improve the prediction of small BER as compared to the Gaussian assumption, because the measured moments would have to be known very precisely [176, 177].

Despite the limitations of the Q-factor method, it is widely used to predict the BER. The method is so popular that even *measured* BER (not necessarily for OOK signals only) are sometimes expressed as a Q-factor according to Eq. (8.37). However, for advanced formats like M-ary quadrature amplitude modulation (QAM), a different metric has to be employed.

A.4.4 Error Vector Magnitude and BER for *M*–ary QAM Signals

Advanced modulation formats such as *M*-ary QAM encode data in amplitude and phase of the optical carrier. Its complex amplitude "vector" E is received with a coherent receiver and described by M points (symbols) in a complex constellation plane. Examples for 16QAM, 32QAM and 64QAM constellations are shown in the upper insets of Fig. A.9 (from left to right). The actually received signal vector E_r deviates by an error vector E_{err} from the ideal transmitted vector E_t. Fig. A.8(a) depicts the ideal constellation points for a 16QAM signal. In Fig. A.8(b) a simulated noisy constellation is shown.

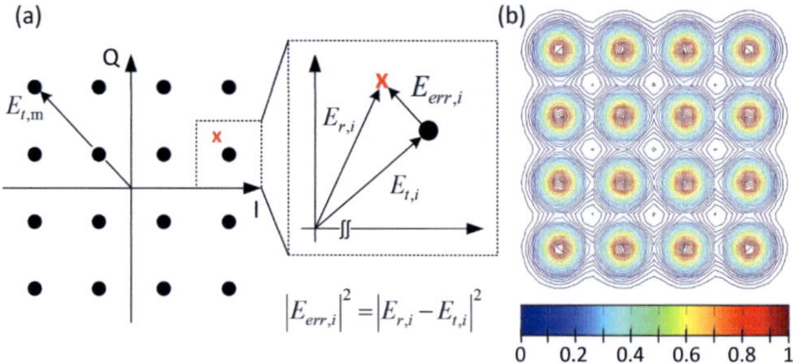

Fig. A.8: Constellation diagram and error vector for a 16QAM signal. (a) Ideal constellation diagram with an actually transmitted value X. The blow-up illustrates the definition of the ith error vector $E_{err,i}$ in relation to the actually received signal vector $E_{r,i}$ and the vector $E_{t,i}$ of the transmitted signal. (b) Simulated constellation diagram with white Gaussian noise, EVM = 15 %. We use the color coding to indicate the reception probability of transmitted symbols. [Figure reprinted from [85] © 2012cIEEE]

The error vector magnitude (EVM) is defined by a root mean square σ_{err} of the various E_{err} for N_S randomly transmitted symbols and embraces all (linear and nonlinear) impairments. We subscript the EVM depending on the normalization, which is either the maximum field $|E_{t,\,m}|$ in the constellation, or the average field $|E_{t,\,a}|$ resulting from summing the powers of all M possible symbols. The ratio k depends on the modulation format (binary and quadrature PSK, 8PSK and xQAM). We find [85] (OSNR means the signal-to-noise power ratio measured in the optical signal bandwidth, not in a reference bandwidth of 1 nm (12.5 GHz at $\lambda = 1.55$ µm). *Erratum* [147]: In Eq. (4), replace $\sqrt{2}$ by 1):

$$\mathrm{EVM}_m = \frac{\sigma_{err}}{|E_{t,\,m}|}, \qquad \sigma_{err}^2 = \frac{1}{N_S}\sum_{i=1}^{N_S} |E_{err,\,i}|^2, \qquad E_{err,\,i} = E_{r,\,i} - E_{t,\,i},$$

$$\mathrm{EVM}_a = k\,\mathrm{EVM}_m, \qquad |E_{t,\,a}|^2 = \frac{1}{M}\sum_{i=1}^{M} |E_{t,\,i}|^2, \qquad k = \frac{|E_{t,\,m}|}{|E_{t,\,a}|}. \tag{8.39}$$

	B/Q/8PSK	16QAM	32QAM	64QAM
k^2	1	9 / 5	17 / 10	7 / 3

The EVM$_m$ from Eq. (8.39) can then be estimated from measured OSNR. Basic assumptions are AWGN, nondata-aided reception and quadratically arranged xQAM constellations where $\log_2(x)$ is even. We find from [65, 133, 186]:

$$\text{EVM}_a = \left[\frac{1}{\text{OSNR}} - \sqrt{\frac{96/\pi}{(M-1)\text{OSNR}}} \sum_{i=1}^{\sqrt{M}-1} \gamma_i \, e^{-\alpha_i} + \frac{12}{M-1} \sum_{i=1}^{\sqrt{M}-1} \gamma_i \beta_i \, \text{erfc}\left(\sqrt{\alpha_i}\right) \right]^{1/2},$$

$$\alpha_i = \frac{3\beta_i^2 \, \text{OSNR}}{2(M-1)}, \quad \beta_i = 2i-1, \quad \gamma_i = 1 - \frac{i}{\sqrt{M}}. \tag{8.40}$$

As mentioned after Eq. (8.38), the relation for the first term $\text{OSNR} \approx \text{EVM}_a^{-2} = (|E_{t,a}|/\sigma_{\text{err}})^2$ in Eq. (8.40) resembles $\gamma \approx Q^2$, which holds for direct reception and electronic amplifier noise only. The remaining terms account for nondata-aided reception and disappear for large OSNR. For $M \gg 2$ only the first few terms in the summation need to be considered. If for data-aided reception the EVM is known, the BER can be approximated by [65, 85, 133, 186]

$$\text{BER} = \frac{1-M^{-1/2}}{\frac{1}{2}\log_2 M} \text{erfc}\left[\frac{3/2}{(M-1)\text{EVM}_a^2} \right]^{1/2}. \tag{8.41}$$

A.4.5   Experimental Results

We measure OSNR, EVM and BER in a setup with a software-defined real-time multi-format transmitter [3], Fig. A.9. We employ 6 modulation formats — B/Q/8PSK and 16/32/64QAM — at symbol rates of 20 GBd and 25 GBd, which are encoded on an external cavity laser (ECL) at 1550 nm. The modulated carrier is kept at a fixed average power and combined with variable ASE noise for changing the OSNR. After an erbium-doped fibre amplifier (EDFA) the OSNR is measured by an OSA. The symbol rate of 25 GBd occupies an optical bandwidth of 25 GHz equal to 2 B_{ref}, so that $\text{OSNR} = \text{OSNR}_{\text{ref}}$, a relation which approximately holds also for 20 GBd. An Agilent modulation analyzer (OMA) decodes the signal and measures EVM and BER. The insets of Fig. A.9 show ASE spectrum and 25 GBd constellations for our best OSNR.

In Fig. A.10(a), measured EVM$_m$ for various measured OSNR values are displayed. Closed (open) symbols represent symbol rates of 20 (25) GBd. Solid lines were calculated with Eq. (8.39), (8.40). For OSNR < 20 dB the prediction agrees with all measurements. Constellations for xQAM can be recovered only if OSNR > 12 dB. For OSNR > 20 dB the constant electronic receiver noise dominates, and an OSNR change does not influence the measured EVM. With smaller electronic noise in systems having smaller symbol rate (smaller bandwidth), the EVM floor would shift to higher OSNR. Figure A.5(b) shows measured BER values as a function of EVM, again for symbol rates of 20 (25) GBd, marked with closed (open) symbols. The solid lines represent Eq. (8.41), the dashed lines result from simulations. While measurement and simulation are based on nondata-aided reception, Eq. (8.41) holds for data-aided reception. Still, measurement, analytical estimate and simulations coincide for $\text{BER} \leq 10^{-2}$. The 32QAM constellation is not quadratic (so Eq. (8.40), (8.41) should not be applied). Still, the constellation is nearly quadratic, and hence the estimation quality remains reasonably good. The plots also reveal that higher-order formats are more noise sensitive, Eq. (8.40), (8.41). — For measuring BER we use a $2^{15} - 1$ pseudo random binary sequence. The number of compared bits and the number of recorded errors were chosen according to the statistical reasoning described in [175-177].

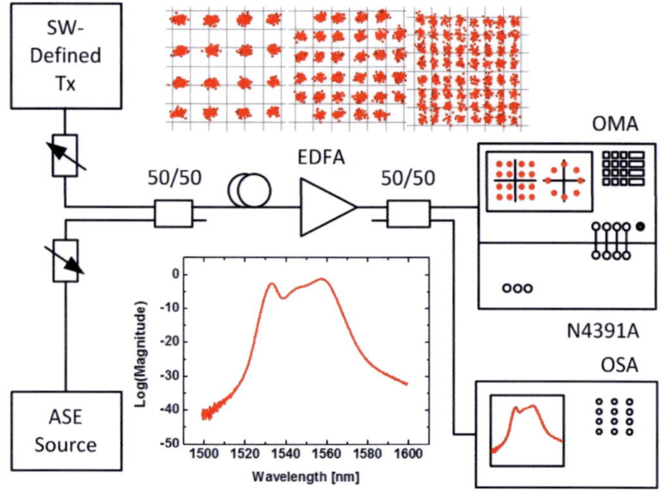

Fig. A.9: Setup for BER and EVM measurements. Software-defined transmitter Tx [4] with ASE-adjustable OSNR (inset: spectrum) and erbium-doped fibre amplifier EDFA. OSNR measured with optical spectrum analyzer (OSA). Modulation decoded by Agilent optical modulation analyzer (OMA). [Figure reprinted from [85] © 2012cIEEE]

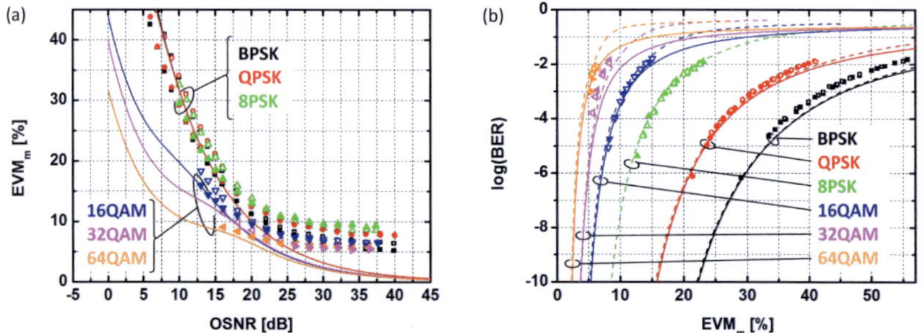

Fig. A.10: EVM$_m$(OSNR) and BER(EVM$_m$) for various modulation formats. Symbols: Measurements for symbol rate 20 GBd (filled) or 25 GBd (open). (a) EVM$_m$ = EVM$_a$ / k calculated from measured OSNR by Eq. (8.40) (solid lines). EVM floor due to electronic noise of transmitter and receiver. Floor height differences for Q/8PSK and xQAM stem from different factors k. Error floor for BPSK is lower because of Tx-specific properties. (b) BER as measured (symbols), simulated (dashed lines) and calculated (solid lines) for various EVM$_m$. [Figure reprinted from [85] © 2012cIEEE]

A.4.6 Conclusions

Complementing the established Q-factor metric for OOK systems, the EVM is a suitable quality measure for coherent optical transmission systems employing advanced modulation formats. As long as the fluctuations can be described by white additive Gaussian noise, measured or simulated EVM data lead to a reliable estimate of the BER.

A.5 Nyquist Pulse Shaping Details

The following section on the effect of the limited length of Nyquist pulses in digital signal processing, receiver bandwidth and clock recovery is an extract of [6].

Real–time Nyquist pulse generation beyond 100 Gbit/s and its relation to OFDM

R. Schmogrow,   M. Winter,   M. Meyer,   A. Ludwig,   D. Hillerkuss, B. Nebendahl, S. Ben-Ezra, J. Meyer, M. Dreschmann, M. Huebner, J. Becker, C. Koos, W. Freude, and J. Leuthold

Optics Express 20(1), 317-337 (2012). [6]

A.5.1 Oversampled Nyquist pulses with finite–length

An elementary Nyquist shaped impulse with minimum spectral width is a sinc-function infinitely extended in time. Real Nyquist pulses, however, need to be approximated by a finite-length representation. For practical reasons finite impulse response (FIR) filters are used to build the pulse shapes [187, 188]. In addition, for separating the baseband spectrum from its periodic repetitions using realizable filters, oversampling by a factor q (typically $q = 1.2, 2,$ ...) is needed. In this paper we have chosen $q = 2$. This way we will subsequently save FPGA resources since sampling points of adjacent symbols fall onto the same time slot, see Fig. 8 in [6]. However, smaller oversampling factors such as $q = 1.2$ suffice if adequate anti-aliasing filters are available. This would allow us to reduce the required processing speed and DAC sampling rate but comes at the cost of an increased processing complexity.

A suitable FIR filter of order R_O can be constructed by a sequence of R_O delay elements T_{sa} / q with $T_{sa} = 1 / F_s$, and $R_O + 1$ taps in-between. The tapped signals are weighed by a number of R_O so-called filter coefficients h_r and summed up to form the filter output, Fig. A.11. A "one-tap" filter with order $R_O = 0$ reproduces the filter input.

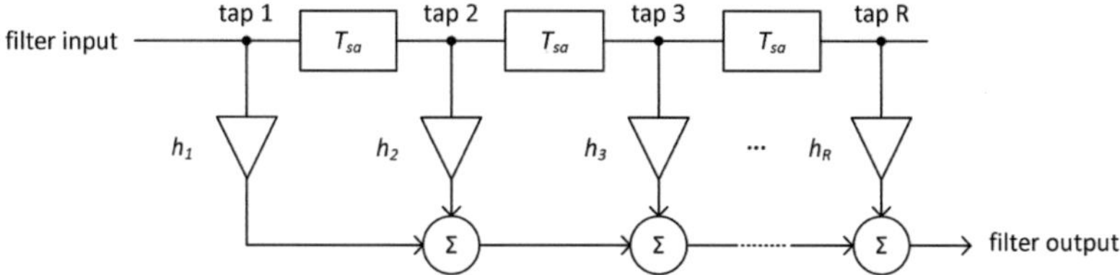

Fig. A.11. Finite impulse response filter (FIR, direct form I) of order R. A series of R delay elements $T_{sa} = T / q$ are located in-between the $R + 1$ taps. Tapped signals are weighed by R filter coefficients h_r and summed to form the filter output.

Signal generation with various FIR filter orders R_O is shown in Fig. A.12. The left column shows the impulse response of each filter. The effective windowing is indicated by a green rectangle. The linearly scaled corresponding transfer functions are seen in the middle column. The right column displays these same transfer functions on a logarithmic scale. The spectra of the single pulses (white lines) are plotted together with simulated data (colored). A two-fold oversampling $q = 2$ is used in this context.

The simulation was performed as follows: A pseudo random binary sequence (PRBS) with a length of $2^{15} - 1$ serves as origin for simulated complex data. As a reference, these complex data c_{ik} modulate NRZ pulses, one of which is displayed in Fig. A.12(a), left column. The linearly scaled sinc-shaped power spectrum of this elementary impulse is seen in Fig. A.12(a), middle column. The logarithm of the same power spectrum is shown as a white line in Fig. A.12(a), right column, together with the ensemble-averaged power spectrum for the simulated data. For all power spectra a possibly existent discrete carrier line is omitted.

Nyquist signals shaped with various FIR filters are depicted in Fig. A.12(b)–(d). The filter order R_O with $R_O + 1$ taps corresponds to the rectangular time window within which the function is defined (left column, green). The convolution of the rectangular spectrum of an infinitely extended temporal sinc-pulse with the sinc-shaped spectrum of the rectangular time window leads to the power spectra depicted in Fig. A.12(b)–(d), middle and right column.

As the filter order R_O increases from $R_O = 16$ to $R_O = 1024$, the spectrum evolves towards an ideal rectangle rect(f / F_s) with a spectral width equal to the Nyquist bandwidth F_s for complex data. Already for $R_O = 32$ a significant increase of the spectral efficiency is to be seen in comparison to NRZ modulation. For $R_O = 1024$ the ideal rectangular spectrum is approximated even more closely. However, due to Gibbs' phenomenon, strong ringing at the steep spectral slopes is to be observed. Non-rectangular window functions like Hann or Hamming windows lead to smoothened spectra and a stronger suppression of the side lobes. However, this advantage comes at the price of a widened spectrum and thus a reduced spectral efficiency.

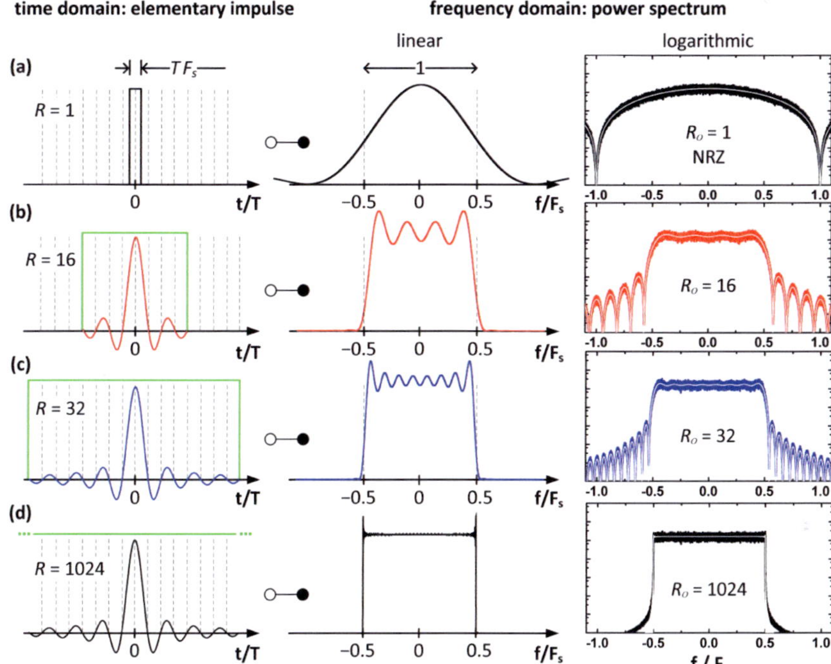

Fig. A.12. Impulse responses and transfer functions of FIR filters with various orders R_O. Left column: Impulse responses. The durations of the impulse response is marked with a rectangular window (green). Middle column: Power spectra on a linear scale. Right column: Power spectra on a logarithmic scale. Colored noisy curves represent the average power spectra for a pulse train with a repetition period $T = 1 / F_s$, which has been encoded with random complex data. The white curves reproduce the power spectra from the middle column. For all power spectra a possibly existent discrete carrier line is omitted. (a) A single NRZ impulse "shaped" by a one-tap "filter" of order $R_O = 0$ leads to a sinc-shaped spectrum. (b) An ideal sinc-impulse is truncated by a rectangular window. The corresponding filter is of order $R_O = 16$. The power spectrum results from the convolution of the rect-shaped spectrum of the sinc-impulse with the sinc-shaped spectrum of the rect-window. (c) An increased filter order of $R_O = 32$ leads to a larger time window, and therefore the resulting spectrum evolves towards an ideal rectangular shape. (d) A filter with very high order $R_O = 1024$ closely approximates a rect-shaped power spectrum. Overshoots and ringing are due to Gibbs' phenomenon. All pulses in these plots have been q-fold oversampled with $q = 2$.

A.5.2   Nyquist pulse reception – required electrical bandwidth and clock phase recovery

Reception of a Nyquist pulse M-ary QAM signal is similar to the reception of a conventional, unfiltered NRZ signal. The complex-modulated optical field is down-converted to the baseband by a coherent receiver (e. g. 90° hybrids with balanced photo-detectors). The electrical signal is then sampled by analog-to-digital converters (ADC) before being processed in the digital domain. Despite all similarities we identified two differences when receiving Nyquist pulses as are discussed in the following.

A.5.2.1 Receiver bandwidth impact

Electrical bandwidth is one major limitation for high-speed signal converters (DACs and ADCs) today. An increase of sampling rate, however, is usually achieved by multiplexing multiple low-speed converters. Hence a high sampling rate is not out of reach, whereas high electrical bandwidth is the much bigger challenge. We want to compare Nyquist signals with standard NRZ signals and the corresponding impact of bandwidth-limited ADCs at the receiver. Therefore we generate two signals, namely a Nyquist signal and an NRZ signal, both with 16QAM modulation at a symbol rate of 14 GBd. Both signals carry the same amount of data (56 Gbit/s). Since the ADCs of our receiver have a fixed analog bandwidth of 32 GHz, we emulate a bandwidth-limited system by applying a digital low-pass filter. The cut-off frequency of this flat-top FIR filter is varied from the minimum Nyquist bandwidth 7 GHz up to 9.33 GHz. A qualitative result is seen in Fig. A.13. The Nyquist signal shows a very clear constellation diagram even at a receiver bandwidth as low as 7 GHz, Fig. A.13(a). The NRZ signal performs poorly under the same conditions, Fig. A.13(b). Increasing the bandwidth to 9.33 GHz improves the signal quality of the NRZ, Fig. A.13(c). Nonetheless the Nyquist signal outperforms the NRZ signal in all cases.

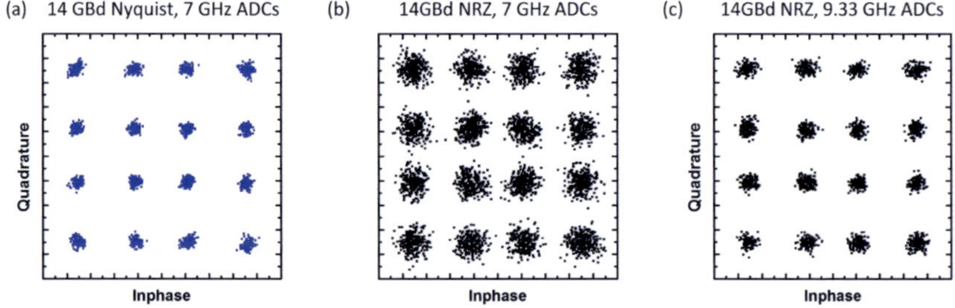

Fig. A.13. Constellation diagrams for Nyquist and NRZ signals received with different bandwidths. (a) A 14 GBd Nyquist signal is well received with a Nyquist bandwidth of 7 GHz. (b) An NRZ of 14 GBd performs poorly with ADCs bandwidth-limited to 7 GHz. (c) Increasing the bandwidth to 9.33 GHz enhances the reception of the NRZ signal. Nevertheless, the Nyquist signal shows best performance.

A.5.2.2 Clock phase recovery

Careful and proper clock recovery is essential for a solid communication link. In order to investigate the influence of the clock phase for close-to-ideal Nyquist signals we replaced the real-time transmitter with an arbitrary waveform generator (AWG). With this transmitter we increased the filter order to 1024, Fig. A.12(d).

For standard NRZ and raised-cosine shaped QAM signals it is common to square the signal [68] and perform a fast Fourier transform (FFT) over several received symbols. The outcome of this procedure is illustrated in Fig. A.14(a). Next to the DC peak we identify two additional peaks with the frequency of the symbol rate. The spectral location of these peaks reveals the symbol rate, and the phase tells the optimum sample time. Squaring the modulated sinc-shaped pulses and performing an FFT leads to Fig. A.14(b). It is obvious that the clock peaks have vanished. Therefore this clock recovery method cannot be applied to sinc-shaped Nyquist signals. Instead we developed an alternative technique to recover the clock phase of a Nyquist signal. In this technique it is sufficient to compute the standard deviation of the modulus of the received Nyquist pulses as a function of the sampling phase. The optimum sampling phase is found when the standard deviation is minimum given that sampling is always done at equivalent positions of subsequent pulses, see Fig. A.14(c). Measured (solid lines) and noiseless signals (dashed lines) agree very well. For QPSK, the so computed standard deviation drops to zero for noiseless signals, see Fig. A.14(c) (QPSK, dashed line). Accumulated noise in measured signals (solid lines) leads to a vertical shift of the curves' minima. Nonetheless, the minimum standard deviation for all signals can be clearly identified.

The algorithm has been tested and works for QPSK to 256QAM and any intermediate M-ary QAM. Since this method neglects the phase of the complex received signal it can be applied prior to carrier phase recovery. Hence standard algorithms for carrier phase recovery can be employed. The influence of a phase error on the signal quality (here represented by EVM) is depicted in Fig. A.14(d). Once the optimum clock phase is found, a feedback control minimizing the signal's EVM is perfectly suited even for real-time systems.

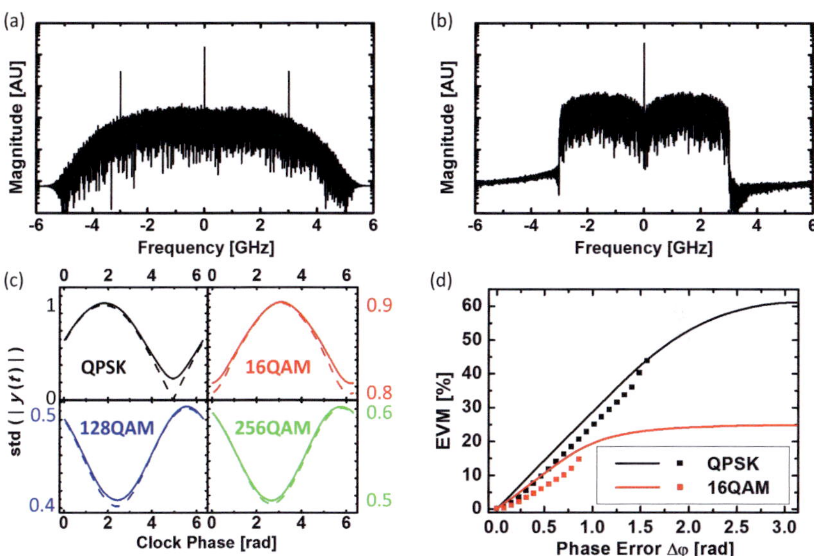

Fig. A.14. Clock phase recovery for M-QAM modulation of raised-cosine and Nyquist pulses, respectively. (a) Spectrum of a squared raised-cosine signal. The clock phase can be extracted from the peaks at the symbol rate. (b) Spectrum of a squared Nyquist sinc-pulse train. The peaks at the symbol rate have vanished. (c) Standard deviation of the modulus of four measured (solid lines) and noiseless (dashed lines) Nyquist signals plotted over the sampling clock phase. Noise leads to a vertical shift of the curves' minima. The initial points of the graphs are arbitrary and depend on the timing of the data acquisition. (d) Dependence of Nyquist signal quality on the clock phase error. Solid line: simulation; squares: measurement.

Accumulated chromatic dispersion (CD) in uncompensated transmission links is usually electrically compensated by a digital filter placed in front of subsequent processing blocks. It is independent of the clock phase recovery described here. Nevertheless, our clock phase recovery algorithm was measured to tolerate a residual dispersion of up to 2400 ps / nm for a 14 GBd QPSK sinc-shaped signal. Furthermore, we found by evaluating transmission experiments, that polarization mode dispersion (PMD) has only negligible influence on the performance of the algorithm.

A.5.3   Spectral efficiency and peak–to–average power ratio

Spectral efficiency (SE) is a major argument for the use of advanced modulation formats in combination with sophisticated multiplexing techniques. Since Nyquist pulses and OFDM signals are closely related, it is interesting to compare the potential SE of both techniques. To this end we compute the spectral width B of the Nyquist pulse up to the first zero outside the main band, which has a width F_s. The same definition is also used for OFDM [61]. The SE results from relating the bitrate of a transmission channel BR (measured in bit/s) to the required transmission bandwidth B, $\mathrm{SE} = \mathrm{BR}\,/\,B$. Information rate and symbol rate are related as follows: For M-ary single-polarization single-carrier Nyquist pulse transmission, the symbol rate is $F_s^{\mathrm{Nyq}} = \mathrm{BR}/\log_2 M$ (in the Nyquist context abbreviated by $F_s = F_s^{\mathrm{Nyq}}$, see Tab. A.1). For single-polarization M-ary OFDM signals with N subcarriers the symbol rate amounts to $F_s^{\mathrm{OFDM}} = \mathrm{BR}/(N \log_2 M)$ (abbreviated in the OFDM context by the same symbol $F_s = F_s^{\mathrm{OFDM}}$, see Tab. A.1).

The transmission bandwidth depends on the respective modulation types and formats. For Nyquist pulses, the spectrum is calculated in the Appendix, Eq. (8.27). Because of the finite length of the actual Nyquist pulses, the spectrum depends on the filter order R_O and the oversampling factor q, see Fig. A.11. For convenience and without loss of generality we choose the spectral symbol $i = 0$ which lies symmetrical to $f = 0$. The spectrum then reads

$$Y^{(0)}\left(f, R_O\right) = \frac{T_S}{\pi}\left[\mathrm{Si}\left(\pi R_O \frac{f + F_s/2}{qF_s} \right) - \mathrm{Si}\left(\pi R_O \frac{f - F_s/2}{qF_s} \right) \right]. \tag{8.42}$$

The function $\mathrm{Si}(z)$ denotes the sine integral [168], see text before Eq. (8.27) in the Appendix. Power spectra computed from Eq. (8.42) closely match the graphs of Fig. A.15 which are obtained by simulations. To determine the bandwidth $B = B^{\mathrm{Nyq}}$, we find the first spectral zeros to the right and to the left of the main band by a numerically exact evaluation of Eq. (5). From these results we extract a simple empirical relation to estimate the SE of digitally generated Nyquist signals:

$$\mathrm{SE}_{\mathrm{Nyquist}} = \begin{cases} \dfrac{\log_2 M}{1 + 2.517/R_O} & \text{for} \quad 1 \le R_O \le 1024, \ q = 2 \\[2ex] \log_2 M & \text{for} \quad R_O > 1024, \ q = 2 \end{cases} \tag{8.43}$$

The resulting spectral efficiency according to Eq. (8.43) is plotted in Fig. A.15(a) (blue line).

For OFDM the SE is influenced by the number of subcarriers N, or in other words by the size of the inverse fast Fourier transform (IFFT) used for signal generation. For our discussion we disregard more advanced OFDM techniques such as a cyclic prefix, guard bands or the introduction of pilot tones that would decrease the SE. The resulting SE then is [61]

$$\mathrm{SE}_{\mathrm{OFDM}} = \frac{\log_2 M}{1 + 1/N}.$$

(8.44)

The normalized spectral efficiencies of OFDM signals are also depicted in Fig. A.15(a) (red line). The SE of both techniques is almost equal.

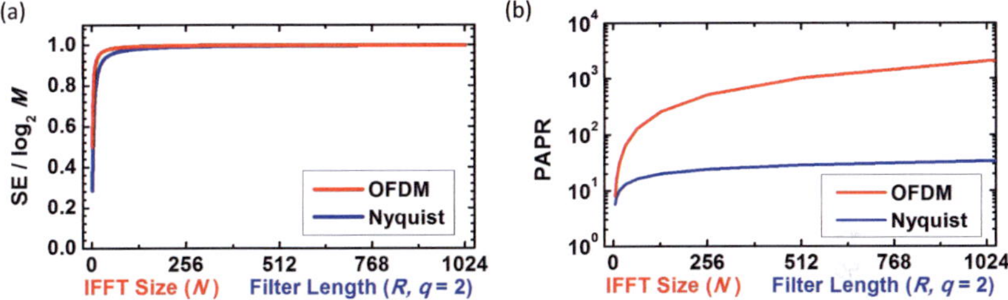

Fig. A.15. Spectral efficiency (SE) and peak-to-average power ratio (PAPR) for Nyquist pulses and OFDM signals as a function of the number R of filter taps and the IFFT size N, respectively. (a) Normalized spectral efficiency (SE) for single-polarization M-ary QAM modulation. Nyquist shaped signals and OFDM signals show almost identical SE. (b) PAPR increases with filter order R for single-carrier Nyquist signals slower than the PAPR for OFDM with the number of subcarriers N.

A major issue that is often referred to reporting on OFDM is the high peak to average power ratio (PAPR) of the time domain signal. This is due to the coherent superposition of multiple sinusoidal carriers that could interfere constructively. As a consequence, high signal amplitudes can occur. In the following we derive PAPR expressions at the transmitter side for Nyquist pulse transmission and for OFDM signaling. At the transmitter, a large PAPR is most critical regarding the rather low resolution of high-speed DACs, the conversion range of which has to be utilized optimally. The PAPR at the receiver end depends heavily on properties of the transmission link like dispersion or nonlinearity tolerance. Therefore, general predictions cannot be made.

To derive an expression for the PAPR in OFDM we need to find the peak power and an expression for the average power. A maximum value can be found as follows: In order to compute the largest possible peak power of an OFDM signal $x(t)$, we assume without loss of generality that the N subcarriers are modulated with a random sequence of real coefficients $c_{ik} = \pm 1$. In this case the maximum amplitude is seen if all N maxima of the temporal sinusoidals happen to add constructively at one point in time, see Fig. 2.22(a) at $t = 0$ and Eq. (8.22) in the Appendix. The average power of such a random OFDM signal is the sum of the average powers of the N orthogonal subcarriers. For arbitrary modulation coefficients c_{ik}, the average power for $c_{ik} = \pm 1$ has to be divided by a format dependent factor k^2 [85, 147]. In real-world OFDM systems the infinitely extended ideal spectrum is narrowed by low-pass filtering, so that the orthogonality relation does not hold any more in the strict sense. Nevertheless, with the orthogonality relation Eq. (8.2) and the power relation Eq. (8.8) we obtain a good approximation of the average power given by Eq. (8.25) with $Q = N$. We thus approximate the PAPR$_{\text{OFDM}}$ by

$$\text{PAPR}_{\text{OFDM}} = \frac{N^2}{\frac{1}{2}N/k^2} = 2k^2 N. \tag{8.45}$$

The result of Eq. (8.45) for $k^2 = 1$ is seen in Fig. A.15(b), red line. With increasing number N of subcarriers the value for PAPR$_{\text{OFDM}}$ increases linearly. However, the probability that an OFDM signal actually has this peak amplitude decreases with the complexity of the M-ary QAM modulation and with the number N of the subcarriers.

For Nyquist signals the PAPR has to be investigated, too, since a superposition of temporally shifted sinc-pulses, see Fig. 2.22(b), also produces high signal amplitudes at certain times. We assume again that the Nyquist pulses are modulated with a random sequence of real coefficients $c_{ik} = \pm 1$. Although the local extrema of a single sinc-impulse are not located at times $t / T_s = -0.5, 0.5, 1.5\dots$, i. e., not in the center of the interval between zeros, it can be shown that the extrema of superimposed Nyquist pulses are located at exactly these times, Eq. (8.13) in the Appendix. For a worst-case consideration all contributions sum up constructively, so in order to obtain the maximum compound signal we sum up the absolute values of sinc-pulses at $t / T_s = 1 / 2$. If the compound Nyquist signal was constructed with infinitely extended sinc-functions, the maximum signal power would not converge when the number of Nyquist pulses increases. Nevertheless, sinc-functions located far away from the time of summation only contribute little to the sum. For a finite approximation of a sinc-impulse as described in Section 4, only R_O / q pulses can contribute. Here the filter order R_O denotes the number of time intervals T_s / q for q-fold oversampling, i. e., R_O stands for the length of the impulse response. We find the maximum power, see Eq. (8.14) in the Appendix with $Q = R_O / q$

$$P_{\text{max}} = \left[\sum_{r=-R/(2q)+1}^{R/(2q)} \left| \text{sinc}\left(\frac{1}{2} - r\right) \right| \right]^2. \tag{8.46}$$

Technically speaking, P_{max} could become arbitrarily large for large filter orders R_O. Yet, while P_{max} increases with R_O, the probability for finding R_O sinc-pulses interfering constructively decreases as well similarly to the OFDM case.

For finalizing the calculation of the PAPR, we need the average power of a single-carrier Nyquist signal $y^{(0)}(t)$ encoded with real coefficients $c_{ik} = \pm 1$. According to [167] we find the average power $\overline{P}$ of an ideal Nyquist signal (see Eq. (8.10) in the Appendix),

$$\overline{P} = \frac{1}{T_s} \int_{-\infty}^{+\infty} \text{sinc}\left(\frac{t}{T_s}\right) dt = 1. \tag{8.47}$$

As for band-limited OFDM spectra, orthogonality is lost for truncated Nyquist sinc-impulses. If Nyquist pulses are generated with a filter of finite (but sufficiently large) order R, orthogonality as implied by Eq. (8.47) is still a good assumption, so that the average power of truncated Nyquist sinc-impulses is close to $\overline{P} = 1$. As before, for arbitrary modulation coefficients c_{ik}, Eq. (8.47) has to be divided by a format dependent factor k^2 [85, 147]. The $\text{PAPR}_{\text{Nyquist}}$ then follows from the ratio of maximum power P_{max} and average power $\overline{P} \approx 1/k^2$,

$$\text{PAPR}_{\text{Nyquist}} = \frac{P_{\text{max}}}{\overline{P}} \approx k^2 \left[\sum_{r=-R/(2q)+1}^{R/(2q)} \left| \text{sinc}\left(\frac{1}{2} - r \right) \right| \right]^2 . \tag{8.48}$$

A detailed mathematical description is given in the Appendix, leading to Eq. (8.20). The PAPR of Nyquist signals from Eq. (8.48) and $k^2 = 1$ is plotted in Fig. A.15(b), blue line. Unlike OFDM signals where the PAPR increases linearly, Eq. (8.45), the PAPR of Nyquist signals does not, due to the temporal decay of its elementary sinc-impulse. However, neither for OFDM nor for single-carrier Nyquist pulses the PAPR converges with increasing IFFT size N or filter order R_O, respectively.

In Nyquist WDM systems, the PAPR could be higher, because multiple spectral symbols separated by at least F_s might add up constructively as well. However, the Nyquist channel spacing is typically in the order of several GHz [13, 68, 69] whereas OFDM carrier spacings are often chosen to be in the MHz [31, 189] range. For large channel spacings as in Nyquist WDM, however, strong signal peaks only occur for very short times, and dispersion causes signal peaks to decay rapidly if large frequency differences are involved. Non-linear effects in WDM systems have been investigated in [190].

A.6 Transmitter Implementation

This section on the implementation of the transmitter is based on [60] and has been extended to include pulse shaping (in Section A.6.4), an example on the data flow in a 16QAM modulation module (end of Section A.6.4), and an example for the compensation of the modulator nonlinearities (end of Section A.6.5).

Implementation of an OFDM Capable Optical Multi-Format Transmitter

R. Schmogrow, Master Thesis No. 803, *Institute of Photonics and Quantum Electronics* (Karlsruhe Institute of Technology, Karlsruhe, 2009). [60]

A.6.1   Digital to Analog Converters

The core component of the transmitter is the high speed digital-to-analog converter (DAC). The fastest DAC that was available with FPGA compatible interfaces was selected for the implementation of the transmitter (Micram – DAC 25). These high-speed digital to analog converters (DAC) can be operated at clock rates up 32 GHz and feature a 4:1 multiplexing circuit for the data inputs (see Fig. A.16). The serial data inputs can therefore operate at up to

8 Gbit/s. For synchronization purposes, helper circuits are located on the DAC. In these circuits, each serial input may be compared to its neighbor to allow for synchronization of the data inputs. Results are written in registers that can be polled by the FPGA. In addition to the data comparison, the sampling point of the DAC can be set in 90° phase steps. This functionality is used when performing phase synchronization of all channels.

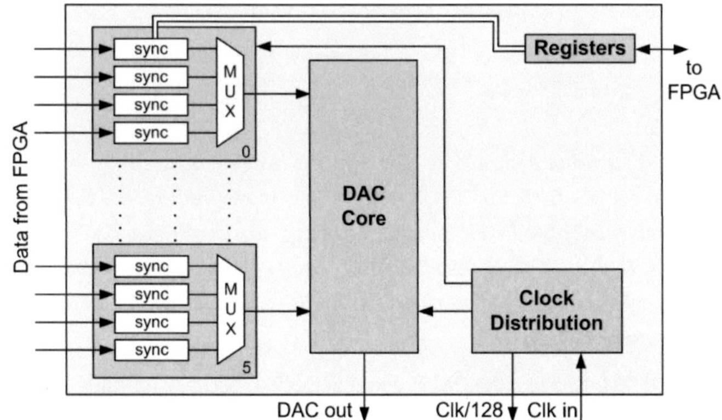

Fig. A.16: Architecture of the digital to analog converter (DAC) used for the implementation of the multi-format transmitter. The data from the FPGA is supplied through high speed interfaces at ¼ of the sampling rate. The data is then fed to synchronization stages and afterwards multiplexed to generate the bit streams at the full sampling rate. These bit streams are then supplied to the DAC core for digital to analog conversion. The DAC is supplied with a sampling clock that is distributed on chip and divided to generate the reference clock (CLK/128) for the FPGA. The registers provide the control functions to the FPGA.

A.6.2 Choice of the FPGA Chip

For the implementation of the transmitter, the choice of FPGA chips was limited. Due to the required speeds of the serial interfaces of the FPGA, only few chips could be chosen, of which only one was available on an evaluation board (Xilinx - ML525) that could be used for the setup. The chip chosen is the Xilinx Virtex 5 - XC5VFX200T with speed grade 2, which is the fastest speed grade available. This chip offers a total number of 24 multi gigabit transceivers (MGT) that are specified for data rates up to 6.5 Gbit/s. As it will be shown later, these chips can be overclocked to achieve the data rates necessary for the DACs.

A.6.3 Multi Gigabit Transmitter Design

The multi gigabit transmitter code design is crucial for successful operation of the transmitter. Due to the 4:1 multiplexing circuit on the DAC, this will result in a sampling clock frequency (CLK) of 25 GHz. The GTX transceivers have a parallel data interface which can be configured for a number of parallel data lines of 32 bits or 40 bit. An interface width of 32 bit is chosen as this fits perfectly to the reference clock (CLK/128 = 195.3 MHz), which is provided by the DAC (Fig. A.16). However, since the high precision clocking inputs of the GTX have to be used for jitter reasons, it is not possible to supply all 24 transmitters using a single clock source input. To overcome this issue the reference clock is split and supplied to two clock inputs of the FPGA with each of the driving a set of GTX transmitters. This fact implies having two clock domains on the FPGA that have the same frequency but not the same phase. For stable operation it has to be taken care of that no errors occur in when data is transferred from one clock domain to the other. Fig. A.17 illustrates the architecture of the user IP-core of the transmitter module.

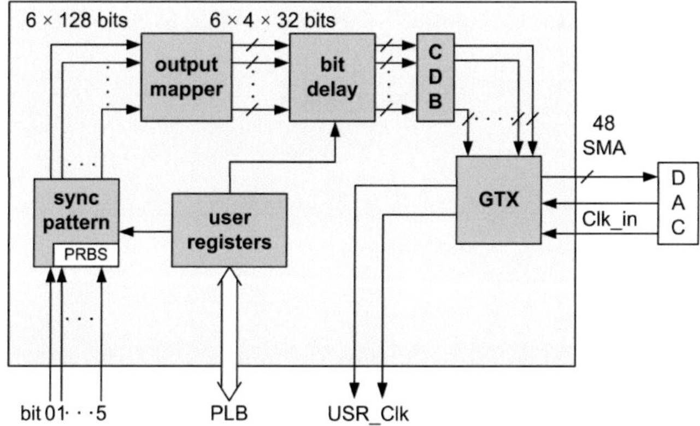

Fig. A.17: Design of the GTX module. Two clock inputs (Clk_in) are provided by the DAC and supplied to the GTX transceivers. An additional clock (DRP Clk) is provided for internal synchronization purposes. The clock is then distributed to additional modules in the chip (USR_Clk). The data bit vectors (bit 0 ... 5) are fed to the sync pattern module, which can either pass the data through, or generate a synchronization patter (e.g. pseudo random bit sequence – PRBS). This data stream is then mapped to the different GTX transceivers and fed to a bit delay for synchronization purposes. After the clock domain bridge (CDB), the data is fed to the GTX transceivers to be sent to the DAC. User registers can be accessed through the processor local bus (PLB) to control the functionality of this module.

The two ports of the reference clock from the DAC are fed to the GTX clock inputs. A separate clock, which is needed to stabilize and synchronize all 24 GTX-transceivers within one FPGA, is supplied through the dynamic reconfiguration port (DRP Clk). This clock is derived from the 50 MHz oscillator which also drives the Power PC core.

At the input six bit vectors of 128 bit have to be provided at every clock cycle of the internal clock (USR_Clk). This specific length originates from the 32 bit data width of the GTX transceivers and the 4:1 multiplexer on the DAC. The sync pattern block is used by the Power PC core to supply the synchronization patterns during the synchronization routine. This block is controlled through software by user registers and is transparent during regular operation. For synchronization purposes the same $2^{31}-1$ pseudo random bit sequence according to [70] will be transmitted on all 24 GTX transmitters simultaneously. The implementation of the bit stream generation is discussed in detail in section A.6.6 yet multiplexing is not necessary as the maximum data rate for synchronization is fixed to 6.25 Gbit/s. The output mapper assigns the components of the input bit vectors to the corresponding transmitters, which are determined by the 4:1 multiplexer and the actual wiring of the DAC and the FPGA. This has the advantage that all prior components of the FPGA design do not have to be aware of the actual multiplexing sequence, thereby reducing complexity outside the actual transmitter module. Also, in case the multiplexing sequence changes, only the mapping has to be readjusted.

Another important block of the transmitter design is the bit delay. For synchronization purposes each GTX transmitter output can be delayed in single bit steps with respect to all other outputs. This block is implemented as 32 bit first-in first-out (FIFO) shift registers with an adjustable output window. The schematic setup of the bit delay block for bit delays up to 32 bits is shown in Fig. A.18.

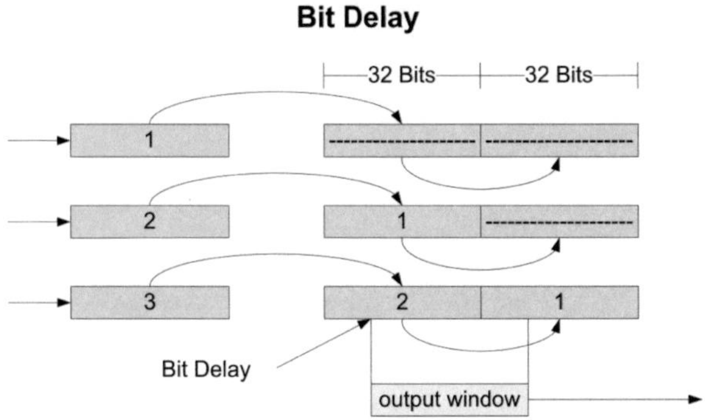

Fig. A.18. Bit Delay FIFO Shift Registers. On each clock cycle, the first-in first-out shift register (FIFO) is shifted by 32 bit. An adjustable output window is used to selects 32 output bits with an arbitrary delay. The maximum delay is determined by the length of the shift register.

At every clock cycle a 32 bit shift is applied to a 64 bit register. A 32 bit output data window masks data to be transmitted by the corresponding GTX transceiver. When adjusting the bit delay value, the offset value for the output window is changed. In that way each of the 24 transmission channels can be delayed by 0 to 31 bit. For the final design, the bit delay was extended to a maximum delay of of 127 bits. The functionality is demonstrated for two bit patterns transmitted at 1.5 Gbit/s (see Fig. A.19). These signal waveforms have been measured using a real-time oscilloscope with 8 GSa/s and an electrical bandwidth of 1.5 GHz. In Fig. A.19, two graphs are displayed, of which the red curve is delayed by two bits with respect black curve. The bit delay block, just like the sync pattern block, is controlled by the Power PC core through user registers. The actual bit delay values are set during synchronization and stay fixed during regular operation.

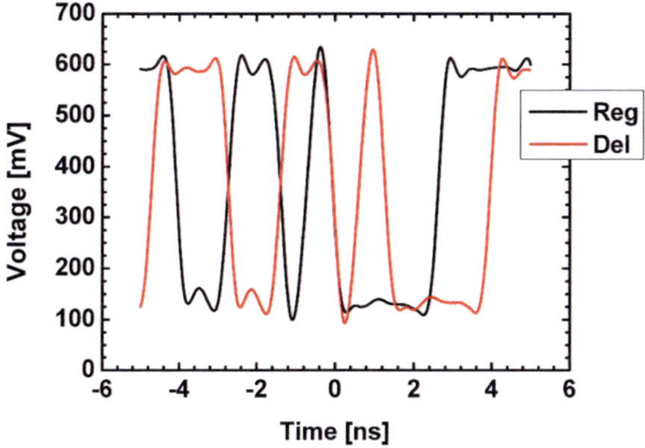

Fig. A.19. PRBS at 1.5 Gbit/s at the output of a GTX transmitter of the FPGA. When comparing the regular output (——) and delayed waveform (——), one can clearly see the functionality of the bit delay block Fig. A.18.

The next block is the clock domain bridge (CDB) that is necessary to overcome data inconsistencies resulting from the two clock domains of the GTX transceivers. These two clock domain clocks have exactly the same frequency as they originate from a common clock. However phase relations cannot be predicted. Therefore so-called asynchronous FIFO registers are inserted between the bit delay block and the actual transmitters. Each transmitter reads out data from the FIFO by the rising edge of its individual clock. On the other side a common clock shared by all modules involved in data generation is used to feed data to the FIFO registers also on the rising edge. Proper operation of these registers as well as the need for solving intersecting clock domain issues has been proven and verified. From this block data is directly fed to the GTX block or GTX wrapper. The GTX wrapper includes stabilization and synchronization of all 24 transceivers. It has been created using the Xilinx IP-Core Generator. With this tool all parameters and settings can be chosen using a wizard which guides the user through all adjustable properties of the transceivers. It also provides an onboard synchronization routine which is described in [191]. Output files of the core generator are in general VHDL files which can be adjusted and inserted in the actual project or design.

A.6.4 Modulation Module

The modulation module is essential for the character of the transmitter. Since the design is held modular and thus very flexible, basically any modulation format can be easily inserted in between the pseudo random bit generation and the transmission block. One of the simplest modulation modules is a sequence memory which holds a user waveform to be repeatedly transmitted by the setup. This memory naturally does not need a pseudo random bit stream generator (PRBS) thus removes complexity from the design. This is to be used for prototyping and initial testing of modulation formats. Transmission of these sequences into the memory located on the FPGA is managed by the control PC and the Power PC core of the

FPGA. Once a waveform has been stored, the transmitter will output the sequence repeatedly. In this operating mode, the transmitter behaves like an arbitrary waveform generator with sampling rates up to 34 GSa/s per channel and two synchronous outputs.

After prototyping, a real-time modulation module can replace offline signal generation. The transmitter can generate any other kind of modulation scheme including on off keying (OOK), phase shift keying (PSK), and Quadrature amplitude modulation (QAM). Fig. A.20 gives an overview of possible configurations of the modulation module.

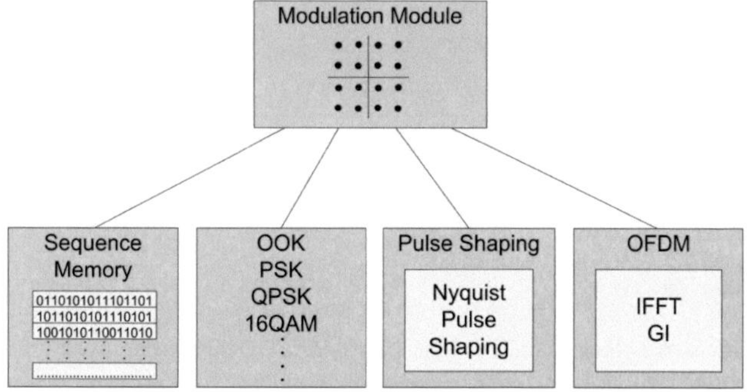

Fig. A.20. Examples for modulation formats that can be implemented in the modulation module. For prototyping, a sequence memory is available to store offline generated waveforms. By changing the modulation module, the transmitter can generate any kind of modulation format like on-off-keying (OOK), phases shift keying (PSK), quadrature phase shift keying (QPSK), quadrature amplitude modulation (QAM). It can also generate perform pulse shaping (e.g. Nyquist pulse shaping) and signals with frequency multiplexing techniques (e.g. orthogonal frequency division multiplexing (OFDM)).

All modulation schemes except for the sequence memory need some sort of bit generation, which is done by a pseudo random bit stream generator as described in A.6.6. As an alternative, an external data source could be supplied. The data generated in the modulation module is sent to GTX module that controls the multi gigabit transmitters (MGT) of the FPGA. The modulation module can be easily extended with more functionalities as displayed in Fig. A.21.

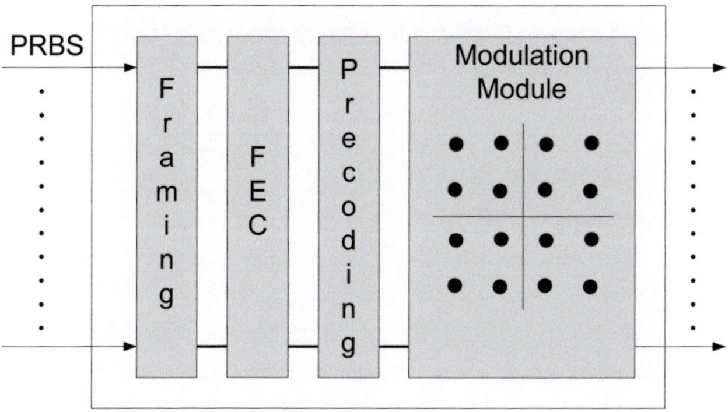

Fig. A.21. Additional features can be included in the modulation module, e.g.: framing, forward error correction (FEC), or any other kind of precoding.

A framing module can take payload data namely pseudo random bit streams and put them into a frame containing additional information such as headers. These can be used e.g. for routing purposes and for control information. Forward error correction (FEC) is another feature which is commonly used in most of today's networks. To avoid that the payload data contains long sequences of ones or zeros in a row or control sequences, encoding techniques such as 8B/10B coding can be applied online. This can be implemented in an encoding module which translates incoming data according to the defined code.

An example for the data flow in a 16QAM modulation module is shown in Fig. A.22. In this module, the bitstream is separated into two bitstreams for in-phase (I) and quadrature (Q) components, which are subsequently encoded in electrical signals with four levels.

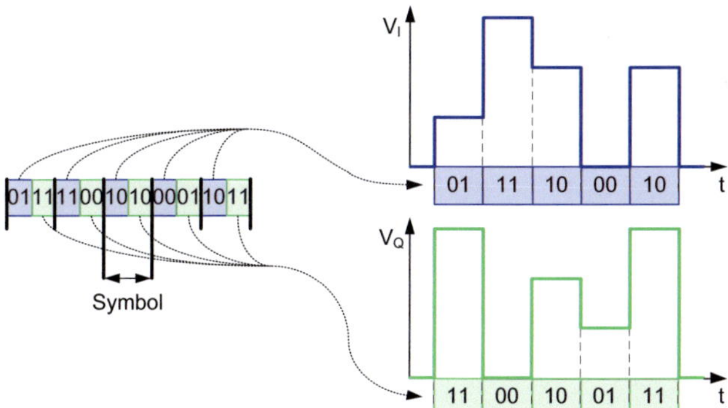

Fig. A.22. Modulation module for 16QAM. In the modulation module for 16QAM, a PRBS at 100 Gbit/s is encoded into two signals with four signal levels each in real-time. Groups of four bits are assigned to the symbols and subsequently demultiplexed to form two data streams of half the data rate for the in-phase and quadrature components. The two bits per symbol now describe two signals with four amplitude levels.

A.6.5   Compensation of the nonlinear transfer function of the IQ modulator

Electrical predistortion is used to compensate the non-linearity of the MZM amplitude transfer function in each arm. The chosen bias for operation is the null point located at a differential phase shift of 180° (known as V_π), which is equal to a phase-shift of 90° in each arm of the modulator. It can be seen from the amplitude transfer function in Fig. A.2 that this bias will yield the best choice to avoid a strong non-linearity. However, when driving the modulator with large amplitudes, the nonlinearities of the transfer function (A.2.1.1) will cause distortions of the signal. Electrical predistortion can overcome these degradations by linearizing the transfer function. When the modulator is properly biased (A.2.2), the amplitude transfer function of each arm will be a true sine wave. As discussed in [192], distorting the signal waveform by an arcsine function will lead to a minimum of non-linear distortions by the Mach Zehnder Modulator (MZM) (8.49).

$$x = \sin\left(\arcsin\left(x\right)\right) \tag{8.49}$$

A graphic interpretation of the above described process is shown in Fig. A.23. The black curve represents the MZM transfer function when biased as discussed. The maximum amplitude of the drive signal is V_π. The actual signal amplitude must be known when applying electrical predistortion. The red curve shows the predistortion function from (8.49). Pre-distorting the electrical signal with an arcsine function will lead to a reduction of the effective resolution of the digital to analog converters (DAC). In the worst case, which is the maximum amplitude, this reduction is about 36 % (8.50).

$$1 - \frac{1}{\pi/2} \approx 0.36 \tag{8.50}$$

To overcome the issue of effective resolution reduction an external predistortion circuit may be used. In that way the whole 6 bit resolution of the DAC boards can be used. Such an external predistortion circuit is proposed in [192] although it does not fulfill the bandwidth requirements for highest data rates. When applying predistortion to the signal the combined transfer function are now linear (blue curve in Fig. A.23).

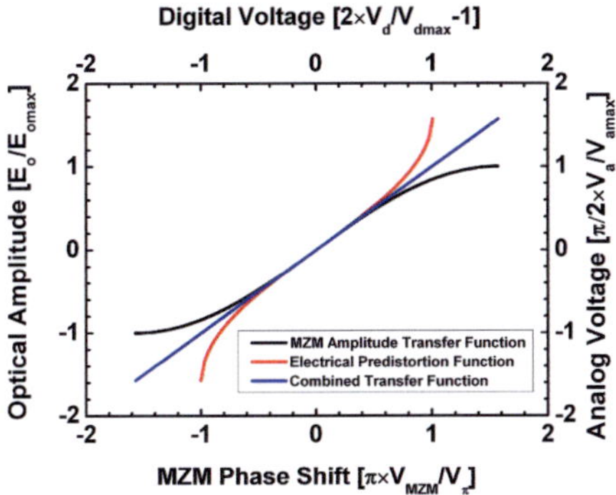

Fig. A.23. Electrical predistortion. To compensate for the nonlinear modulator transfer function (——) the electrical predistortion function (——) is applied. The overall transfer function is now linear (——).

Other applications for electrical predistortion of the modulated signal besides compensating modulator non-linearity is compensating dispersion effects in optical fiber. These techniques have been shown in [81] and can be applied to signals of any modulation format. However, predistortion to compensate for fiber dispersion needs fixed optical channels since dispersion characteristics of the applied fiber span has to be known prior to transmission.

The effect of the nonlinearity of the modulator on a 16 QAM signal is displayed in Fig. A.24. One can see that the constellation points in the constellation diagram are no longer equally spaced (Fig. A.24 right hand side).

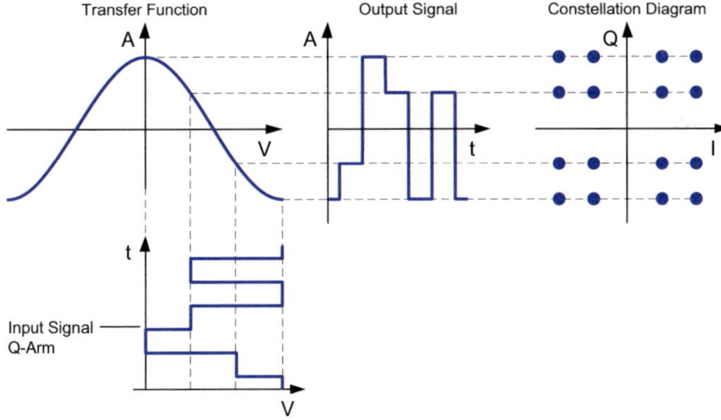

Fig. A.24. Distortion of an ideal 16QAM drive signal. The input signal for the Q-arm has equidistant signal levels. After applying the transfer function of the Mach-Zehnder, the signal levels of the output signal are no longer equidistant. In combination with the I-arm of the IQ-Modulator this leads to constellation points that are not equidistant, limiting the resilience to white noise of the generated 16QAM data signal.

The resilience of the generated data signal to additive noise will therefore be suboptimal. This can be circumvented by pre-compensation of the modulator transfer function as it is illustrated in Fig. A.25.

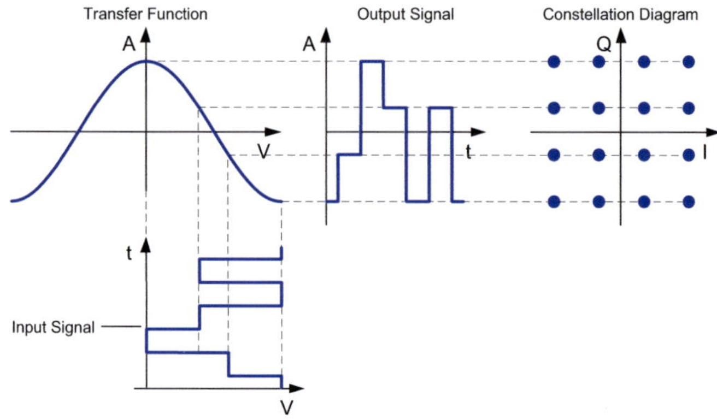

Fig. A.25. An ideal constellation diagram is achieved when a predistortion is applied to the modulator drive signal. The signal levels of the input signal are modified such that the signal levels of the output signal are now equidistant. This leads to an ideal constellation diagram at the transmitter output.

For a 16QAM signal, the pre-compensation has been implemented by slightly modifying the modulation module to generate a signal with non-equidistant signal levels.

A.6.6   Multi Gigabit PRBS Generator

According to the standard described in [70], pseudo random bit sequences (PRBS) are created following well defined computation instructions. One way of creating such sequences is the use of a shifting register and a simple XOR logic gate. Fig. A.26 shows the defining circuit for two PRBS of different lengths. After this length the bit stream is repeated periodically.

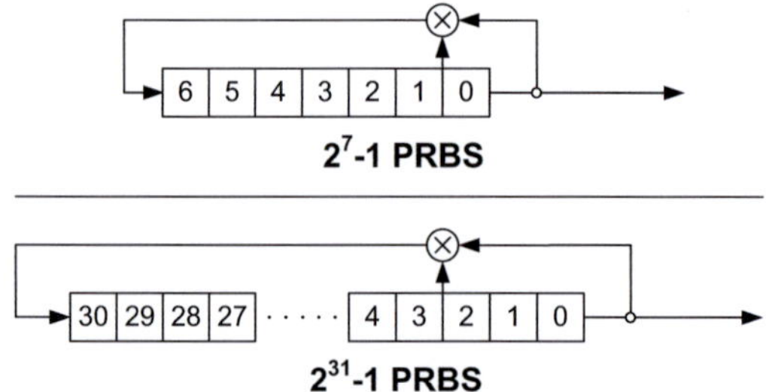

Fig. A.26. Circuits for the generation of pseudo random bit sequences with the lengths of 2^7-1 and 2^{31}-1 bit. The sequence is generated using linear feedback shift registers, where certain parts of the shift register are combined in an XOR ($\otimes$) to be fed back to the input of the shift register.

Using shift registers it is only possible to compute a single bit per clock cycle. However in order to generate multi gigabit streams it is essential to parallelize the bit sequence generation. Running this setup at multiple GHz is not possible, as the maximum internal clock frequency of today's available FPGAs is in the order of 400 MHz. One way to parallelize bit generation is to modify the serial nature of Fig. A.26 to a more parallel process by using a general mathematical description of the given procedure.

$$b_n = b_{n-7} \otimes b_{n-6} \tag{8.51}$$

$$b_n = b_{n-31} \otimes b_{n-28} \tag{8.52}$$

In equations (8.51) and (8.52) b_n is the value of the nth bit which is computed by an XOR logical operation between prior bits of the very same sequence. Equation (8.51) shows the function for generating a PRBS with a length of 2^7-1 whereas (8.52) is related to a PRBS with a length of $2^{31}-1$. Taking advantage of these functions it is convenient to generate up to 16 bits (8.51) respectively 32 bits (8.52) at a single clock cycle. However, this technique is limited as at some point a single XOR gate is no longer sufficient to compute the nth bit. When looking at a PRBS $2^{31}-1$, the generation of the 40th bit e.g. needs the values of the 9th and the 12th bit as inputs. These values are created at the same clock cycle, so (8.52) essentially becomes (8.53).

$$b_{40} = \left(b_{-31} \otimes b_{-28}\right) \otimes \left(b_{-28} \otimes b_{-25}\right) \tag{8.53}$$

At this point already two cascaded logical operations are needed which naturally increases computation time. So the need for very high rates implies using an additional technique which complements the above described parallel bit generation.

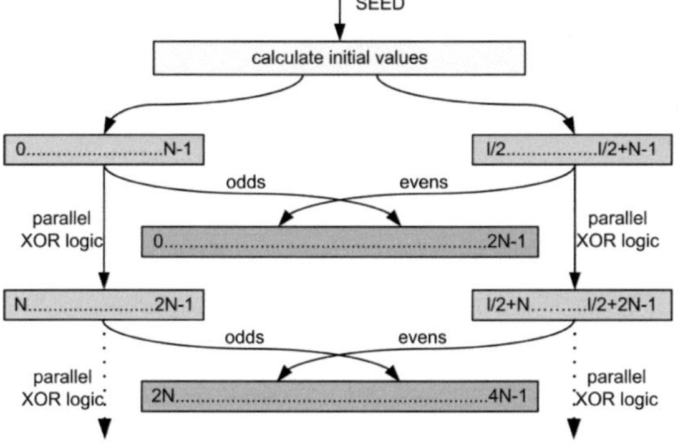

Fig. A.27. Generation of a PRBS using a 2:1 multiplexer and a parallelized PRBS algorithm. The initial values of the first 2N bits of the PRBS are calculated from a seed using the standard shift register method. They are then demultiplexed into two vectors of the length N to form the input of two parallel XOR logic circuits, of which each calculates N bit of the next 2N bit of the PRBS. The newly generated 2N bit of the PRBS are obtained by multiplexing the N bit output sequences of the parallel XOR logic circuits.

Another characteristic of a standard PRBS comes into play. It is well known, that two excerpts of the same PRBS 2^N-1 with $l = 2^N$ that are delayed with respect to each other by $l\,/\,2$, can be multiplexed to form a longer part of the very same bit sequence. The combination of parallel bit generation and multiplexing technique is illustrated for a 2:1 multiplexer (MUX) in Fig. A.27.

Initially a so-called seed has to be fed to the algorithm which calculates two vectors with the length of N parallel bits gained from parallel XOR logic. Both vectors originate from the same PRBS and are delayed by exactly $l\,/\,2$. Multiplexing both vectors in a zipper like fashion, the output vector is of twice the length of the two input vectors. Furthermore the output is identical to the initial PRBS calculated up to two times N. After multiplexing, the algorithm uses parallel bit generation to compute the next two vectors to be multiplexed again. With this algorithm two smaller vectors are calculated simultaneously and a vector twice as long is provided at the output. Functionality of the program has been demonstrated using a Matlab program which compares regular successive generation of a PRBS with the algorithm described above. Essentially, both outputs turned out to be identical. In order to experimentally back up this method, an FPGA design has been implemented which features a 1.5 Gbit/s PRBS using online signal generation. The FPGA provides multi gigabit transceivers (MGT) which are operated at 1.5 Gbit/s each. The MGT takes a vector of 20 bits at a frequency of 75 MHz for transmission. Fig. A.28 shows two PRBS with a length of 2^7-1 that have been generated differently. The black curve has been generated without the multiplexing procedure with exactly 20 bit per clock cycle at 75 MHz while the red curve was created using the algorithm shown in Fig. A.27. For the second curve, the clock frequency for the generation of the sequence was set to 37.5 MHz and the circuit generated 40 bit per cycle. Both output measurements are identical with respect to their binary information and therefore the multiplexing algorithm yields the same result as regular generation, yet giving twice the number of bits per clock cycle.

Further increase of bits per clock cycle implies adding a more complex multiplexing scheme. These more complex algorithms can be implemented by cascading the method described above. For an m:1 multiplexer with $m = 2^n$ (n=1,2,3,4,...) as seen in Fig. A.29, m smaller excerpts of a PRBS are multiplexed to form a longer excerpt of that PRBS.

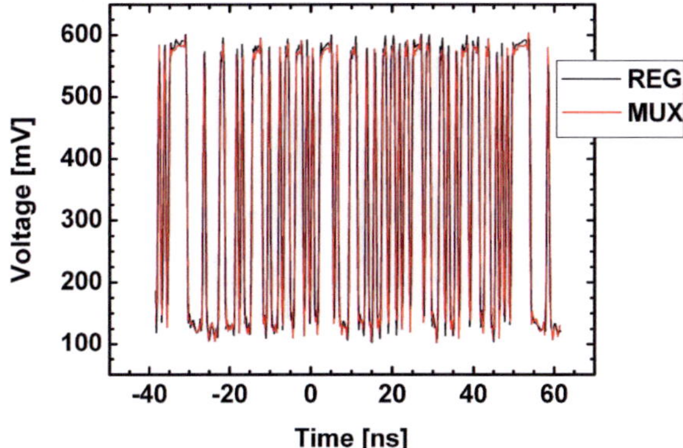

Fig. A.28. Measured PRBS sequences at 1.5 Gbit/s with a length of 2^7-1 that have been generated in the regular way using parallel XOR logic (——) and using multiplexing in addition to generate a larger number of bits per clock cycle (——). It can be seen that the bit sequences are identical.

All excerpts must have a defined offset with respect to the seed and must be multiplexed in the right order. Naturally the number of input vectors to the multiplexer is fixed to a power of two. For the final design an algorithm with 32 bits per cycle and an additional 8:1 multiplexer is used. This way, a number of 256 bit are generated per clock cycle. When operating this PRBS generator at the maximum clock frequency of 390.625 MHz data rates up to 100 Gbit/s can be generated in real-time. These rates could also be achieved storing a sequence in local memory, however, the memory of the FPGA is not large enough to hold the longest PRBS sequences of interest ($2^{31} - 1$), and therefore, online bit generation as described is inevitable.

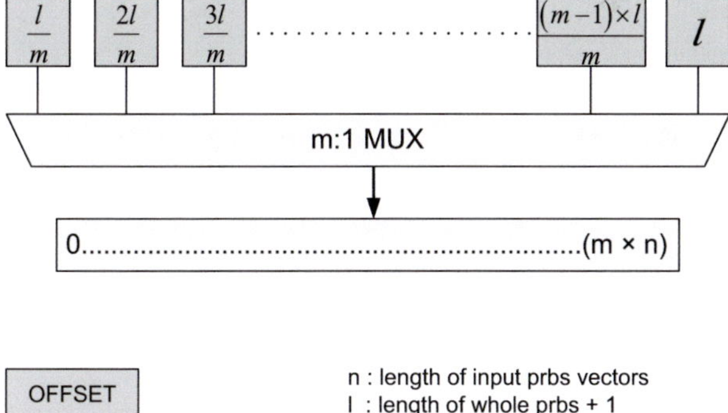

Fig. A.29. Algorithm for the generation of a large number of bits of a PRBS using an m:1 multiplexer. The required offset of the single PRBS generators (blue boxes) is given in this figure. The number of PRBS generators to be multiplexed must be equal to a power of two.

A.7 Waveshaper

This description of the Finisar waveshaper was first published in [166].

Comb Generation for Next Generation Optical Transmission Systems

M. Jordan, Diploma Thesis No. 846, *Institute of Photonics and Quantum Electronics* (Karlsruhe Institute of Technology, Karlsruhe, 2012) [166].

The waveshaper (WS) is a fully programmable filter with 1:4 demultiplexing or 4:1 multiplexing capabilities. It is able to create a wavelength dependent attenuation and phase characteristic. Furthermore, it offers the possibility to switch the input signal to different output ports depending on the wavelength. This way a waveshaper acts as a wavelength selective switch (WSS) with the possibility to modify the spectral shape for each channel.

The setup of such a ROADM is shown in Fig. A.30. The spectral components of the input signal into the Waveshaper have to be separated in space. This is done in a dispersive element which is build up from a combination of prism and a grating Fig. A.30 (6). By refractive (3) and mirroring (4) optics the spectral components are imaged on the horizontal axis of the modulating element (7). This modulating element is a liquid crystal on silicon (LCoS) micro display.

On the LCoS display the different spectral components can be addressed by pixel columns. In the waveshaper used in the experiments, each pixel column covers a spectral range of about 7 GHz. For each of those pixel columns the signal can be set in phase and amplitude. The resulting transfer function of the waveshaper is approximately the convolution of the Gaussian beam caused by the optics and rectangular transfer function of the pixel columns. Thus the spectral resolution of the waveshaper is limited to about 7 GHz.

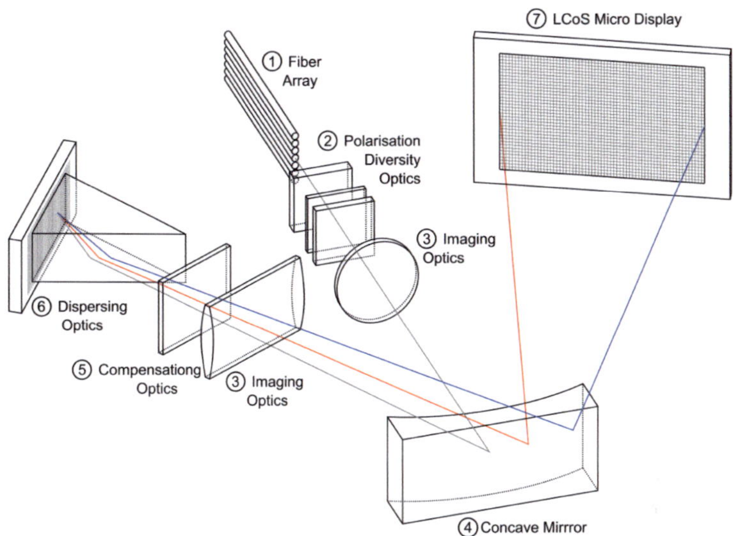

Fig. A.30: Optical scheme of an LCoS ROADM [166]: Light which enters WS through the fiber (1) gets reflected by a mirror (4) and divided in its spectral components (5). The dispersed beam gets reflected again and projected onto the micro display (6). The light is reflected by the display and routed backwards to the fiber array. By using some imaging optics (3) the beam is spread over the display height and refocused in the output fiber. Polarization diversity optics (2) separates two orthogonal polarizations and turns one by 90° to overcome the polarization sensitivity of the LCoS display.

In vertical direction the beam is stretched by imaging optics to cover a large number of pixel rows. This way each pixel column covers a certain spectral range. By applying a saw tooth structure to a pixel column it is possible to build a Fresnel lens on the reflecting surface (Fig. A.31). Through this, the incident wave is reflected at the required angle towards the desired output port. The beam has to cover a large number of pixel rows to route the beam precisely to the desired output port. By a slight detuning of the vertical saw tooth profile it is possible to reduce the coupling efficiency into the output fiber. This way, the attenuation for each wavelength can be adjusted.

By shifting the phase of a whole pixel column relative to the neighboring column it is possible to control the phase of the different wavelength.

As the overall transfer function depends also on the beam waist of the imaging system, neighboring wavelengths cannot be affected completely independent. When implementing a large attenuation step, a significant amount of light is leaking into neighboring wavelengths. High phase steps will affect the neighboring wavelength and lead to a reduced power caused by destructive interference.

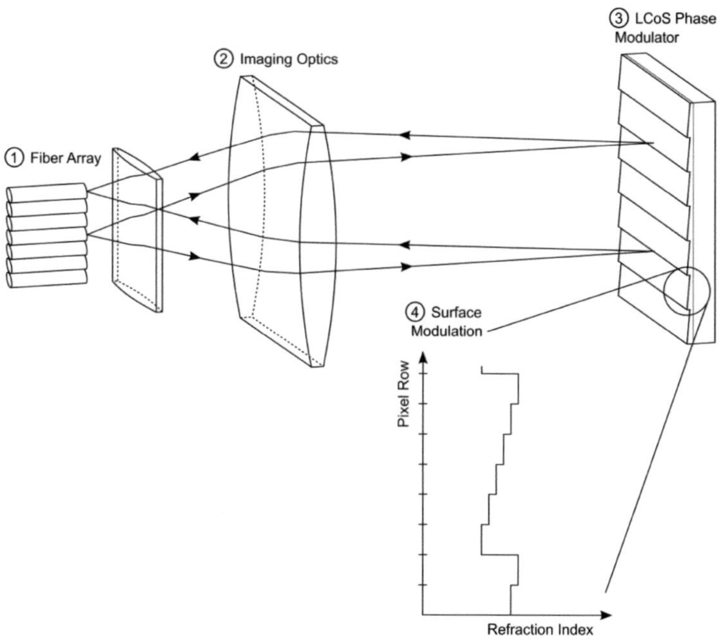

Fig. A.31: Port routing of an LCoS ROADM [166]: Light coming through a fiber (1) is spread over a large amount of pixel rows on an LCoS micro display (3). Controlling the pixels LC refraction index changes the phase of the reflected light. By apply a linear tilt (4) in the phase the direction of the reflected light can be controlled. To overcome a limited control range of the phase, a phase reset after 2π is performed. The surface of the LCoS is plane. The physical saw tooth surface shown in the figure (3) illustrates the effect via an equivalent phase shift by a physical profile.

The LCoS display is operated as a phase modulator of the reflected light. By applying a voltage to a liquid crystal its refraction index changes. This way it is possible to modify the optical path length and thus the phase of the signal. Fig. A.32 shows the buildup of an LCoS structure. On a silicon substrate there are CMOS structures to control the voltage of the LCoS cell. The system is operated in reflection. To improve this reflection an additional reflective coating is put on the substrate. On top of the reflective coating the liquid crystal is placed. In the presence of an electric field the crystals get aligned and the whole layer becomes birefringent. An alignment layer improves the crystals alignment. To build up a electric field over the LC a second electrode is needed. An indium tin oxide (ITO) layer serves as transparent electrode and closes the electric circuit over the LC. The whole system is closed and protected by a cover glass.

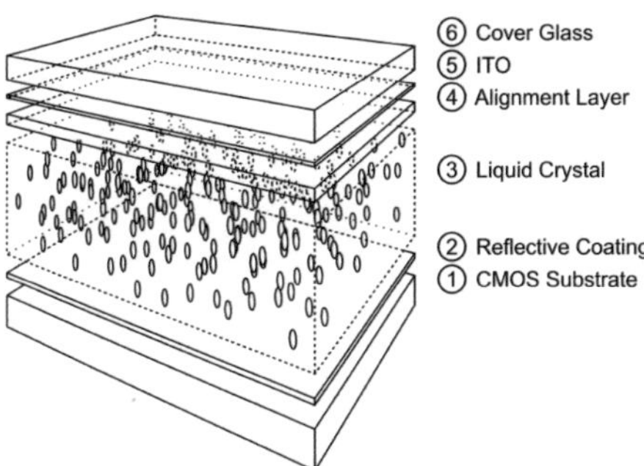

Fig. A.32: LCoS buildup [166]: A CMOS structure (1) contains a pixel matrix of electrodes with driver electronics. A reflection layer (2) improves the reflection of the incident light. Liquid crystal (3) is responsible for the modulation. Depending of the applied electric field the crystals are aligned. The aligned crystal becomes birefringent and changes its refractive index. This results in a phase change of reflected light. To keep the crystals properly aligned, an alignment layer (4) is used. The transparent indium tin oxide (ITO) layer (5) acts as the second electrode to establish the electric field. The system is closed and protected by a cover glass (6).

The phase shift is highly polarization dependent because the refractive index change is introduced by birefringence of the liquid crystals. Therefore a polarization diversity scheme is required (Fig. A.30 (2)). The incoming signal is split into orthogonal polarizations. One of them is rotated by 90 degree and shifted vertically. This way both polarizations are processed in parallel and recombined before they enter the output ports.

Appendix B: References

[1]   D. Hillerkuss, R. Schmogrow, M. Huebner, M. Winter, B. Nebendahl, J. Becker, W. Freude, and J. Leuthold, "Software-defined multi-format transmitter with real-time signal processing for up to 160 Gbit/s," in *Signal Processing in Photonic Communications* (Optical Society of America, 2010), paper SPTuC4.

[2]   R. H. Stolen and C. Lin, "Self-phase-modulation in silica optical fibers," Phys. Rev. A **17**(4), 1448-1453 (1978).

[3]   R. Schmogrow, D. Hillerkuss, M. Dreschmann, M. Huebner, M. Winter, J. Meyer, B. Nebendahl, C. Koos, J. Becker, W. Freude, and J. Leuthold, "Real-time software-defined multiformat transmitter generating 64QAM at 28 GBd," IEEE Photon. Technol. Lett. **22**(21), 1601-1603 (2010).

[4]   R. Schmogrow, M. Winter, B. Nebendahl, D. Hillerkuss, J. Meyer, M. Dreschmann, M. Huebner, J. Becker, C. Koos, W. Freude, and J. Leuthold, "101.5 Gbit/s real-time OFDM transmitter with 16QAM modulated subcarriers," in *Optical Fiber Communication Conference* (2011), paper OWE5.

[5]   R. Schmogrow, M. Meyer, S. Wolf, B. Nebendahl, D. Hillerkuss, B. Baeuerle, M. Dreschmann, J. Meyer, M. Huebner, J. Becker, C. Koos, W. Freude, and J. Leuthold, "150 Gbit/s real-time Nyquist pulse transmission over 150 km SSMF enhanced by DSP with dynamic precision," in *Optical Fiber Communication Conference* (2012), paper OM2A.6.

[6]   R. Schmogrow, M. Winter, M. Meyer, A. Ludwig, D. Hillerkuss, B. Nebendahl, S. Ben-Ezra, J. Meyer, M. Dreschmann, M. Huebner, J. Becker, C. Koos, W. Freude, and J. Leuthold, "Real-time Nyquist pulse generation beyond 100 Gbit/s and its relation to OFDM," Opt. Express **20**(1), 317-337 (2012).

[7]   D. Hillerkuss, T. Schellinger, R. Schmogrow, M. Winter, T. Vallaitis, R. Bonk, A. Marculescu, J. Li, M. Dreschmann, J. Meyer, S. Ben-Ezra, N. Narkiss, B. Nebendahl, F. Parmigiani, P. Petropoulos, B. Resan, K. Weingarten, T. Ellermeyer, J. Lutz, M. Moller, M. Huebner, J. Becker, C. Koos, W. Freude, and J. Leuthold, "Single source optical OFDM transmitter and optical FFT receiver demonstrated at line rates of 5.4 and 10.8 Tbit/s," in *Optical Fiber Communication Conference* (2010), paper PDPC1.

[8]   D. Hillerkuss, R. Schmogrow, T. Schellinger, M. Jordan, M. Winter, G. Huber, T. Vallaitis, R. Bonk, P. Kleinow, F. Frey, M. Roeger, S. Koenig, A. Ludwig, A. Marculescu, J. Li, M. Hoh, M. Dreschmann, J. Meyer, S. Ben-Ezra, N. Narkiss, B. Nebendahl, F. Parmigiani, P. Petropoulos, B. Resan, A. Oehler, K. Weingarten, T. Ellermeyer, J. Lutz, M. Moeller, M. Huebner, J. Becker, C. Koos, W. Freude, and J. Leuthold, "26 Tbit s^{-1} line-rate super-channel transmission utilizing all-optical fast Fourier transform processing," Nature Photon. **5**(6), 364-371 (2011).

[9]   D. Hillerkuss, A. Marculescu, J. Li, M. Teschke, G. Sigurdsson, K. Worms, S. Ben-Ezra, N. Narkiss, W. Freude, and J. Leuthold, "Novel optical fast Fourier transform scheme enabling real-time OFDM processing at 392 Gbit/s and beyond," in *Optical Fiber Communication Conference* (2010), paper OWW3.

[10]   D. Hillerkuss, M. Winter, M. Teschke, A. Marculescu, J. Li, G. Sigurdsson, K. Worms, S. Ben Ezra, N. Narkiss, W. Freude, and J. Leuthold, "Simple all-optical FFT scheme enabling Tbit/s real-time signal processing," Opt. Express **18**(9), 9324-9340 (2010).

[11]   "OFC/NFOEC – Corning Outstanding Student Paper Competition,",
http://www.ofcnfoec.org/Home/Submit-Papers/Corning-Oustanding-Student-Paper-
Award.aspx, Accessed 13/05/2012, 2012.

[12]   D. Hillerkuss, R. Schmogrow, M. Meyer, S. Wolf, M. Jordan, P. Kleinow, N.
Lindenmann, P. Schindler, A. Melikyan, X. Yang, S. Ben-Ezra, B. Nebendahl, M.
Dreschmann, J. Meyer, F. Parmigiani, P. Petropoulos, B. Resan, A. Oehler, K.
Weingarten, L. Altenhain, T. Ellermeyer, M. Moeller, M. Huebner, J. Becker, C.
Koos, W. Freude, and J. Leuthold, "Single-laser 32.5 Tbit/s Nyquist WDM
transmission," J. Opt. Commun. Netw. **4**(10), 715-723 (2012).

[13]   D. Hillerkuss, R. Schmogrow, M. Meyer, S. Wolf, M. Jordan, P. Kleinow, N.
Lindenmann, P. Schindler, A. Melikyan, X. Yang, S. Ben-Ezra, B. Nebendahl, M.
Dreschmann, J. Meyer, F. Parmigiani, P. Petropoulos, B. Resan, A. Oehler, K.
Weingarten, L. Altenhain, T. Ellermeyer, M. Moeller, M. Huebner, J. Becker, C.
Koos, W. Freude, and J. Leuthold, "Single-laser 32.5 Tbit/s Nyquist WDM
transmission," ArXiv e-prints arXiv:1203.2516v1 (2012),

http://arxiv.org/abs/1203.2516v1.

[14]   D. Hillerkuss, R. Schmogrow, C. Koos, W. Freude, J. Leuthold, B. Nebendahl, and T.
Dippon, "Real-time Nyquist Pulse Transmission," ITG Symposium on Photonic
Networks (2012).

[15]   D. Hillerkuss, R. Schmogrow, M. Meyer, S. Wolf, M. Jordan, P. Kleinow, N.
Lindenmann, P. C. Schindler, A. Melikyan, M. Dreschmann, J. Meyer, J. Becker, C.
Koos, W. Freude, J. Leuthold, X. Yang, F. Parmigiani, P. Petropoulos, S. Ben-Ezra, B.
Nebendahl, B. Resan, A. Oehler, K. Weingarten, L. Altenhain, T. Ellermeyer, M.
Moeller, and M. Huebner, "Single-laser 32.5 Tbit/s Nyquist-WDM transmission over
227 km with real-time Nyquist pulse shaping," in *Photonics in Switching* (2012), pp.
Th-S13-I06.

[16]   K. Coffman and A. Odlyzko, "The size and growth rate of the internet," First Monday
3(10) (1998).

[17]   "Cisco visual networking index: Forecast and methodology, 2010–2015,"   Cisco
Systems Inc., Cisco VNI, 2011 (2011),

http://www.cisco.com/en/US/solutions/collateral/ns341/ns525/ns537/ns705/ns827/whi
te_paper_c11-481360_ns827_Networking_Solutions_White_Paper.html.

[18]   "Corning® SMF-28® ULL optical fiber datasheet," (Corning Inc., 2012),
http://www.corning.com/opticalfiber/products/SMF-28_ULL_fiber.aspx.

[19]   "2011 optical cable demand rose 9%, passed 200 M F-km," (CRU Group, 2012),
http://cruonline.crugroup.com/WireandCable/Marketanalysis/OpticalFibreandFibreOp
ticCableMonitor/tabid/115/Default.aspx, Accessed March 1, 2012, 2012.

[20]   Q. Li, D. Fortusini, S. D. Benjamin, G. Qi, P. V. Kelkar, and V. L. da Silva, "Gain-
flattened, extended L-band (1570-1620 nm), high power, low noise erbium-doped
fiber amplifiers," in *Optical Fiber Communication Conference* (2002), pp. 459-461.

[21]   R. J. Mears, L. Reekie, I. M. Jauncey, and D. N. Payne, "Low-noise erbium-doped
fibre amplifier operating at 1.54µm," Electron. Lett. **23**(19), 1026-1028 (1987).

[22]   "Understanding fiber optics: Attenuation,"   Furukawa Electric North America, Inc.,
(2007),

http://www.ofsoptics.com/resources/Understanding-Attenuation.pdf.

[23]   P. J. Winzer, "Modulation and multiplexing in optical communications," in
Conference on Lasers and Electro-Optics (CLEO) (2009), paper CTuL3.

[24]   A. H. Gnauck, G. Raybon, S. Chandrasekhar, J. Leuthold, C. Doerr, L. Stulz, A. Agarwal, S. Banerjee, D. Grosz, S. Hunsche, A. Kung, A. Marhelyuk, D. Maywar, M. Movassaghi, X. Liu, C. Xu, X. Wei, and D. M. Gill, "2.5 Tb/s (64×42.7 Gb/s) transmission over 40×100 km NZDSF using RZ-DPSK format and all-Raman-amplified spans," in *Optical Fiber Communication Conference* (2002), pp. FC2-1-FC2-3.

[25]   R. A. Griffin, R. I. Johnstone, R. G. Walker, J. Hall, S. D. Wadsworth, K. Berry, A. C. Carter, M. J. Wale, J. Hughes, P. A. Jerram, and N. J. Parsons, "10 Gb/s optical differential quadrature phase shift key (DQPSK) transmission using GaAs/AlGaAs integration," in *Optical Fiber Communication Conference* (2002), pp. FD6-1-FD6-3.

[26]   C. Wree, N. Hecker-Denschlag, E. Gottwald, P. Krummrich, J. Leibrich, E. D. Schmidt, B. Lankl, and W. Rosenkranz, "High spectral efficiency 1.6-b/s/Hz transmission (8 x 40 Gb/s with a 25-GHz grid) over 200-km SSMF using RZ-DQPSK and polarization multiplexing," IEEE Photon. Technol. Lett. **15**(9), 1303-1305 (2003).

[27]   S. Okamoto, K. Toyoda, T. Omiya, K. Kasai, M. Yoshida, and M. Nakazawa, "512 QAM (54 Gbit/s) coherent optical transmission over 150 km with an optical bandwidth of 4.1 GHz," in *European Conference on Optical Communication* (2010), pp. PD2-3.

[28]   M. Dreschmann, J. Meyer, M. Hubner, R. Schmogrow, D. Hillerkuss, J. Becker, J. Leuthold, and W. Freude, "Implementation of an ultra-high speed 256-point FFT for Xilinx Virtex-6 devices," in *IEEE International Conference on Industrial Informatics (INDIN)* (2011), pp. 829-834.

[29]   W. Freude, R. Schmogrow, D. Hillerkuss, J. Meyer, M. Dreschmann, B. Nebendahl, M. Huebner, J. Becker, C. Koos, and J. Leuthold, "Reconfigurable optical transmitters and receivers," in *Photonics West*, G. Li, and D. S. Jager, eds. (SPIE, 2012), pp. 82840A-82848.

[30]   W. Freude, R. Schmogrow, B. Nebendahl, D. Hillerkuss, J. Meyer, M. Dreschmann, M. Huebner, J. Becker, C. Koos, and J. Leuthold, "Software-defined optical transmission," in *International Conference on Transparent Optical Networks* (2011), paper TU.D1.1.

[31]   R. Schmogrow, M. Winter, D. Hillerkuss, B. Nebendahl, S. Ben-Ezra, J. Meyer, M. Dreschmann, M. Huebner, J. Becker, C. Koos, W. Freude, and J. Leuthold, "Real-time OFDM transmitter beyond 100 Gbit/s," Opt. Express **19**(13), 12740-12749 (2011).

[32]   R. Schmogrow, M. Winter, M. Meyer, D. Hillerkuss, B. Nebendahl, J. Meyer, M. Dreschmann, M. Huebner, J. Becker, C. Koos, W. Freude, and J. Leuthold, "Real-time Nyquist pulse modulation transmitter generating rectangular shaped spectra of 112 Gbit/s 16QAM signals," in *Signal Processing in Photonic Communications* (Optical Society of America, 2011), paper SPMA5.

[33]   T. Richter, E. Palushani, C. Schmidt-Langhorst, M. Nölle, R. Ludwig, and C. Schubert, "Single wavelength channel 10.2 Tb/s TDM-data capacity using 16-QAM and coherent detection," in *Optical Fiber Communication Conference* (2011), paper PDPA9.

[34]   A. Sano, T. Kobayashi, S. Yamanaka, A. Matsuura, H. Kawakami, Y. Miyamoto, K. Ishihara, and H. Masuda, "102.3-Tb/s (224 x 548-Gb/s) C- and extended L-band all-Raman transmission over 240 km using PDM-64QAM single carrier FDM with digital pilot tone," in *Optical Fiber Communication Conference* (Optical Society of America, 2012), paper PDP5C.3.

[35]   D. Qian, M.-F. Huang, E. Ip, Y.-K. Huang, Y. Shao, J. Hu, and T. Wang, "101.7-Tb/s (370×294-Gb/s) PDM-128QAM-OFDM transmission over 3×55-km SSMF using

pilot-based phase noise mitigation," in *Optical Fiber Communication Conference* (2011), paper PDPB5.

[36]   R. Chang, "Synthesis of band-limited orthogonal signals for multichannel data transmission," Bell System Technical Journal **45**(1775-1796 (1966).

[37]   W. Shieh and I. Djordjevic, *OFDM for optical communications* (Elsevier Academic Press, Amsterdam Heidelberg [u.a.], 2010).

[38]   J. Yu, Z. Dong, X. Xiao, Y. Xia, S. Shi, C. Ge, W. Zhou, N. Chi, and Y. Shao, "Generation, transmission and coherent detection of 11.2 Tb/s (112x100Gb/s) single source optical OFDM superchannel," in *Optical Fiber Communication Conference* (2011), paper PDPA6.

[39]   S. L. Jansen, I. Morita, T. C. Schenk, and H. Tanaka, "Long-haul transmission of 16×52.5 Gbits/s polarization-division- multiplexed OFDM enabled by MIMO processing (Invited)," J. Opt. Netw. **7**(2), 173-182 (2008).

[40]   Y. Ma, Q. Yang, Y. Tang, S. Chen, and W. Shieh, "1-Tb/s single-channel coherent optical OFDM transmission over 600-km SSMF fiber with subwavelength bandwidth access," Opt. Express **17**(11), 9421-9427 (2009).

[41]   G. Bosco, V. Curri, A. Carena, P. Poggiolini, and F. Forghieri, "On the performance of Nyquist-WDM terabit superchannels based on PM-BPSK, PM-QPSK, PM-8QAM or PM-16QAM subcarriers," J. Lightwave Technol. **29**(1), 53-61 (2011).

[42]   M. Yan, Z. Tao, W. Yan, L. Li, T. Hoshida, and J. C. Rasmussen, "Experimental comparison of no-guard-interval-OFDM and Nyquist-WDM superchannels," in *Optical Fiber Communication Conference* (Optical Society of America, 2012), paper OTh1B.2.

[43]   J. Yu, Z. Dong, and N. Chi, "30-Tb/s (3×12.84-Tb/s) signal transmission over 320km using PDM 64-QAM modulation," in *Optical Fiber Communication Conference* (Optical Society of America, 2012), paper OM2A.4.

[44]   R. Cigliutti, A. Nespola, D. Zeolla, G. Bosco, A. Carena, V. Curri, F. Forghieri, Y. Yamamoto, T. Sasaki, and P. Poggiolini, "Ultra-long-haul transmission of 16x112 Gb/s spectrally-engineered DAC-generated Nyquist-WDM PM-16QAM channels with 1.05x(symbol-rate) frequency spacing," in *Optical Fiber Communication Conference* (Optical Society of America, 2012), paper OTh3A.3.

[45]   A. D. Ellis and F. C. G. Gunning, "Spectral density enhancement using coherent WDM," IEEE Photon. Technol. Lett. **17**(2), 504-506 (2005).

[46]   T. Sakamoto, T. Kawanishi, and M. Izutsu, "Widely wavelength-tunable ultra-flat frequency comb generation using conventional dual-drive Mach-Zehnder modulator," Electron. Lett. **43**(19), 1039-1040 (2007).

[47]   T. Kawanishi, S. Oikawa, K. Higuma, and M. Izutsu, "Electrically tunable delay line using an optical single-side-band modulator," IEEE Photon. Technol. Lett. **14**(10), 1454-1456 (2002).

[48]   S. Chandrasekhar, L. Xiang, B. Zhu, and D. W. Peckham, "Transmission of a 1.2-Tb/s 24-carrier no-guard-interval coherent OFDM superchannel over 7200-km of ultra-large-area fiber," in *European Conference on Optical Communication* (2009), paper PD2.6.

[49]   X. Yang, D. Richardson, and P. Petropoulos, "Nonlinear generation of ultra-flat broadened spectrum based on adaptive pulse shaping," in *European Conference on Optical Communication* (Optical Society of America, 2011), paper We.7.A.2.

[50]   C. Campopiano and B. Glazer, "A coherent digital amplitude and phase modulation scheme," IRE Transactions on Communications Systems **10**(1), 90-95 (1962).

[51]   G. Grau and W. Freude, *Optische Nachrichtentechnik: Eine Einführung* (Springer Berlin, Heidelberg, 1991).

[52]   P. A. Humblet and M. Azizoglu, "On the bit error rate of lightwave systems with optical amplifiers," J. Lightwave Technol. **9**(11), 1576-1582 (1991).

[53]   S. R. Chinn, D. M. Boroson, and J. C. Livas, "Sensitivity of optically preamplified DPSK receivers with Fabry-Perot filters," J. Lightwave Technol. **14**(3), 370-376 (1996).

[54]   J. Li, K. Worms, D. Hillerkuss, B. Richter, R. Maestle, W. Freude, and J. Leuthold, "Tunable free space optical delay interferometer for demodulation of differential phase shift keying signals," in *Optical Fiber Communication Conference* (2010), paper JWA24.

[55]   J. Li, K. Worms, R. Maestle, D. Hillerkuss, W. Freude, and J. Leuthold, "Free-space optical delay interferometer with tunable delay and phase," Opt. Express **19**(12), 11654-11666 (2011).

[56]   J. Li, C. Schmidt-Langhorst, R. Schmogrow, D. Hillerkuss, M. Lauermann, M. Winter, K. Worms, C. Schubert, C. Koos, W. Freude, and J. Leuthold, "Self-coherent receiver for PolMUX coherent signals," in *Optical Fiber Communication Conference* (Optical Society of America, 2011), paper OWV5.

[57]   J. Li, K. Worms, A. Marculescu, D. Hillerkuss, S. Ben-Ezra, W. Freude, and J. Leuthold, "Optical vector signal analyzer based on differential detection with inphase and quadrature phase control," in *Signal Processing in Photonic Communications* (Optical Society of America, 2010), paper SPTuB3.

[58]   J. Li, K. Worms, P. Vorreau, D. Hillerkuss, A. Ludwig, R. Maestle, S. Schule, U. Hollenbach, J. Mohr, W. Freude, and J. Leuthold, "Optical vector signal analyzer based on differential direct detection," in *LEOS Annual Meeting* (2009), paper TuA4.

[59]   S. Tsukamoto, D. S. Ly-Gagnon, K. Katoh, and K. Kikuchi, "Coherent demodulation of 40-Gbit/s polarization-multiplexed QPSK signals with 16-GHz spacing after 200-km transmission," in *Optical Fiber Communication Conference* (2005), paper 3 pp. Vol. 5.

[60]   R. Schmogrow, "Implementation of an OFDM capable optical multiformat transmitter," Master Thesis No. 803, *Institute of Photonics and Quantum Electronics (IPQ)* (University of Karlsruhe (TH), Karlsruhe, 2009).

[61]   W. Shieh, H. Bao, and Y. Tang, "Coherent optical OFDM: theory and design," Opt. Express **16**(2), 841-859 (2008).

[62]   L. Xiang, S. Chandrasekhar, Z. Benyuan, P. J. Winzer, A. H. Gnauck, and D. W. Peckham, "448-Gb/s reduced-guard-interval CO-OFDM transmission over 2000 km of ultra-large-area fiber and five 80-GHz-grid ROADMs," J. Lightwave Technol. **29**(4), 483-490 (2011).

[63]   M. Nakazawa, S. Okamoto, T. Omiya, K. Kasai, and M. Yoshida, "256 QAM (64 Gbit/s) coherent optical transmission over 160 km with an optical bandwidth of 5.4 GHz," in *Optical Fiber Communication Conference* (2010), paper OMJ5.

[64]   R. J. Essiambre, G. Kramer, P. J. Winzer, G. J. Foschini, and B. Goebel, "Capacity limits of optical fiber networks," J. Lightwave Technol. **28**(4), 662-701 (2010).

[65]   A. D. Ellis, Z. Jian, and D. Cotter, "Approaching the non-linear shannon limit," J. Lightwave Technol. **28**(4), 423-433 (2010).

[66]   M. Sjödin, P. Johannisson, H. Wymeersch, P. A. Andrekson, and M. Karlsson, "Comparison of polarization-switched QPSK and polarization-multiplexed QPSK at 30 Gbit/s," Opt. Express **19**(8), 7839-7846 (2011).

[67]   H. Nyquist, "Certain topics in telegraph transmission theory," American Institute of Electrical Engineers, Transactions of the **47**(2), 617-644 (1928).

[68]   Z. Dong, J. Yu, H.-C. Chien, N. Chi, L. Chen, and G.-K. Chang, "Ultra-dense WDM-PON delivering carrier-centralized Nyquist-WDM uplink with digital coherent detection," Opt. Express **19**(12), 11100-11105 (2011).

[69]   X. Zhou, L. E. Nelson, P. Magill, B. Zhu, and D. W. Peckham, "8x450-Gb/s,50-GHz-spaced,PDM-32QAM transmission over 400km and one 50GHz-grid ROADM," in *Optical Fiber Communication Conference* (2011), paper PDPB3.

[70]   ITU-T, "O.150 – Series O: Specifications of measuring equipment," (1996), pp. 3-5.

[71]   K. B. Howell, *Principles of Fourier analysis* (Chapman & Hall/CRC, 2001).

[72]   J. W. Cooley and J. W. Tukey, "An algorithm for the machine calculation of complex Fourier series," Math. Comp. **19**(297--301 (1965).

[73]   C. E. Shannon, "Communication in the presence of noise," Proceedings of the IRE **37**(1), 10-21 (1949).

[74]   J. P. Costas, "Phase-shift radio teletype," Proceedings of the IRE **45**(1), 16-20 (1957).

[75]   P. J. Winzer and R.-J. Essiambre, "Advanced modulation formats for high-capacity optical transport networks," J. Lightwave Technol. **24**(12), 4711-4728 (2006).

[76]   A. H. Ballard, "A new multiplex technique for communication systems," Power Apparatus and Systems, IEEE Transactions on **PAS-85**(10), 1054-1059 (1966).

[77]   A. Sano, T. Kobayashi, K. Ishihara, H. Masuda, S. Yamamoto, K. Mori, E. Yamazaki, E. Yoshida, Y. Miyamoto, T. Yamada, and H. Yamazaki, "240-Gb/s polarization-multiplexed 64-QAM modulation and blind detection using PLC-LN hybrid integrated modulator and digital coherent receiver," in *European Conference on Optical Communication* (2009), paper PD2.4.

[78]   W. Shieh and C. Athaudage, "Coherent optical orthogonal frequency division multiplexing," Electron. Lett. **42**(10), 587-589 (2006).

[79]   R. I. Killey, P. M. Watts, V. Mikhailov, M. Glick, and P. Bayvel, "Electronic dispersion compensation by signal predistortion using digital processing and a dual-drive Mach-Zehnder Modulator," IEEE Photon. Technol. Lett. **17**(3), 714-716 (2005).

[80]   P. M. Watts, R. Waegemans, Y. Benlachtar, V. Mikhailov, P. Bayvel, and R. I. Killey, "10.7 Gb/s transmission over 1200 km of standardsingle-mode fiber by electronic predistortionusing FPGA-based real-time digital signalprocessing," Opt. Express **16**(16), 12171-12180 (2008).

[81]   R. Waegemans, S. Herbst, L. Holbein, P. Watts, P. Bayvel, C. Fürst, and R. I. Killey, "10.7 Gb/s electronic predistortion transmitter using commercial FPGAs and D/A converters implementing real-time DSP for chromatic dispersion and SPM compensation," Opt. Express **17**(10), 8630-8640 (2009).

[82]   P. J. Winzer, A. H. Gnauck, S. Chandrasekhar, S. Draving, J. Evangelista, and B. Zhu, "Generation and 1,200-km transmission of 448-Gb/s ETDM 56-Gbaud PDM 16-QAM using a single I/Q modulator," in *European Conference on Optical Communication* (2010), paper PD2.2.

[83]   Y. Benlachtar, P. M. Watts, R. Bouziane, P. Milder, D. Rangaraj, A. Cartolano, R. Koutsoyannis, J. C. Hoe, M. Püschel, M. Glick, and R. I. Killey, "Generation of optical OFDM signals using 21.4 GS/s real time digital signal processing," Opt. Express **17**(20), 17658-17668 (2009).

[84]   J. Leuthold, M. Winter, W. Freude, C. Koos, D. Hillerkuss, T. Schellinger, R. Schmogrow, T. Vallaitis, R. Bonk, A. Marculescu, J. Li, M. Dreschmann, J. Meyer,

M. Huebner, J. Becker, S. Ben-Ezra, N. Narkiss, B. Nebendahl, F. Parmigiani, P. Petropoulos, B. Resan, A. Oehler, K. Weingarten, T. Ellermeyer, J. Lutz, and M. Möller, "All-optical FTT signal processing of a 10.8 Tb/s single channel OFDM signal," in *Photonics in Switching* (Optical Society of America, 2010), paper PWC1.

[85]   R. Schmogrow, B. Nebendahl, M. Winter, A. Josten, D. Hillerkuss, S. Koenig, J. Meyer, M. Dreschmann, M. Huebner, C. Koos, J. Becker, W. Freude, and J. Leuthold, "Error vector magnitude as a performance measure for advanced modulation formats," IEEE Photon. Technol. Lett. **24**(1), 61-63 (2012).

[86]   L. Xiang, S. Chandrasekhar, Z. Benyuan, and D. W. Peckham, "Efficient digital coherent detection of A 1.2-Tb/s 24-carrier no-guard-interval CO-OFDM signal by simultaneously detecting multiple carriers per sampling," in *Optical Fiber Communication Conference* (2010), paper OWO2.

[87]   G. Bosco, A. Carena, V. Curri, P. Poggiolini, and F. Forghieri, "Performance limits of Nyquist-WDM and CO-OFDM in high-speed PM-QPSK systems," IEEE Photon. Technol. Lett. **22**(15), 1129-1131 (2010).

[88]   D. Hillerkuss, T. Schellinger, M. Jordan, C. Weimann, F. Parmigiani, B. Resan, K. Weingarten, S. Ben-Ezra, B. Nebendahl, C. Koos, W. Freude, and J. Leuthold, "High-Quality Optical Frequency Comb by Spectral Slicing of Spectra Broadened by SPM," IEEE Photonics Journal **5**(5), 7201011-7201011 (2013).

[89]   J. L. Hall, "Optical frequency measurement: 40 years of technology revolutions," IEEE J. Sel. Topics Quantum Electron. **6**(6), 1136-1144 (2000).

[90]   T. Udem, R. Holzwarth, and T. W. Hansch, "Optical frequency metrology," Nature **416**(6877), 233-237 (2002).

[91]   S. T. Cundiff and A. M. Weiner, "Optical arbitrary waveform generation," Nature Photon. **4**(11), 760-766 (2010).

[92]   T. M. Fortier, M. S. Kirchner, F. Quinlan, J. Taylor, J. C. Bergquist, T. Rosenband, N. Lemke, A. Ludlow, Y. Jiang, C. W. Oates, and S. A. Diddams, "Generation of ultrastable microwaves via optical frequency division," Nature Photon. **5**(7), 425-429 (2011).

[93]   P. J. Delfyett, S. Gee, Myoung-Taek Choi, H. Izadpanah, Wangkuen Lee, S. Ozharar, F. Quinlan, and T. Yilmaz, "Optical frequency combs from semiconductor lasers and applications in ultrawideband signal processing and communications," J. Lightwave Technol. **24**(7), 2701-2719 (2006).

[94]   N. K. Fontaine, G. Raybon, B. Guan, A. L. Adamiecki, P. Winzer, R. Ryf, A. Konczykowska, F. Jorge, J.-Y. Dupuy, L. L. Buhl, S. Chandrasekhar, R. Delbue, P. Pupalaikis, and A. Sureka, "228-GHz Coherent Receiver using Digital Optical Bandwidth Interleaving and Reception of 214-GBd (856-Gb/s) PDM-QPSK," in *European Conference on Optical Communication* (Optical Society of America, 2012), paper Th.3.A.1.

[95]   J. Pfeifle, C. Weimann, F. Bach, J. Riemensberger, K. Hartinger, D. Hillerkuss, M. Jordan, R. Holtzwarth, T. J. Kippenberg, J. Leuthold, W. Freude, and C. Koos, "Microresonator-based optical frequency combs for high-bitrate WDM data transmission," in *Optical Fiber Communication Conference* (Optical Society of America, 2012), paper OW1C.4.

[96]   L. Boivin, M. C. Nuss, W. H. Knox, and J. B. Stark, "206-channel chirped-pulse wavelength-division multiplexed transmitter," Electron. Lett. **33**(10), 827-829 (1997).

[97]   S. Kray, F. Spöler, T. Hellerer, and H. Kurz, "Electronically controlled coherent linear optical sampling for optical coherence tomography," Opt. Express **18**(10), 9976 (2010).

[98]   I. Coddington, W. C. Swann, L. Nenadovic, and N. R. Newbury, "Rapid and precise absolute distance measurements at long range," Nature Photon. **3**(6), 351-356 (2009).

[99]   A. Mishra, R. Schmogrow, I. Tomkos, D. Hillerkuss, C. Koos, W. Freude, and J. Leuthold, "Flexible RF-Based Comb Generator," (2013).

[100]   P. Del'Haye, T. Herr, E. Gavartin, M. L. Gorodetsky, R. Holzwarth, and T. J. Kippenberg, "Octave spanning tunable frequency comb from a microresonator," Phys. Rev. Lett. **107**(6), 063901 (2011).

[101]   M. A. Foster, J. S. Levy, O. Kuzucu, K. Saha, M. Lipson, and A. L. Gaeta, "Silicon-based monolithic optical frequency comb source," Opt. Express **19**(15), 14233-14239 (2011).

[102]   G. Agrawal, *Nonlinear Fiber Optics* (Elsevier Science, 2001).

[103]   E. L. Goldstein, L. Eskildsen, and A. F. Elrefaie, "Performance implications of component crosstalk in transparent lightwave networks," IEEE Photon. Technol. Lett. **6**(5), 657-660 (1994).

[104]   L. Richter, H. Mandelberg, M. Kruger, and P. McGrath, "Linewidth determination from self-heterodyne measurements with subcoherence delay times," IEEE J. Quantum Electron. **22**(11), 2070-2074 (1986).

[105]   D. Derickson and C. Hentschel, *Fiber optic test and measurement* (Prentice Hall PTR New Jersey, 1998).

[106]   M. E. Marhic, "Discrete Fourier transforms by single-mode star networks," Opt. Lett. **12**(1), 63-65 (1987).

[107]   A. E. Siegman, "Fiber Fourier optics," Opt. Lett. **26**(16), 1215-1217 (2001).

[108]   A. E. Siegman, "Fiber Fourier optics: previous publication," Opt. Lett. **27**(6), 381-381 (2002).

[109]   S. Kodama, T. Ito, N. Watanabe, S. Kondo, H. Takeuchi, H. Ito, and T. Ishibashi, "2.3 picoseconds optical gate monolithically integrating photodiode and electroabsorption modulator," Electron. Lett. **37**(19), 1185-1186 (2001).

[110]   H. Sanjoh, E. Yamada, and Y. Yoshikuni, "Optical orthogonal frequency division multiplexing using frequency/time domain filtering for high spectral efficiency up to 1 bit/s/Hz," in *Optical Fiber Communication Conference* (2002), paper ThD1.

[111]   A. J. Lowery, "Design of arrayed-waveguide grating routers for use as optical OFDM demultiplexers," Opt. Express **18**(13), 14129-14143 (2010).

[112]   C. K. Madsen and J. H. Zhao, *Optical filter design and analysis: A signal processing approach* (John Wiley & Sons, Inc., 1999).

[113]   B. H. Verbeek, C. H. Henry, N. A. Olsson, K. J. Orlowsky, R. F. Kazarinov, and B. H. Johnson, "Integrated four-channel Mach-Zehnder multi/demultiplexer fabricated with phosphorous doped SiO2 waveguides on Si," J. Lightwave Technol. **6**(6), 1011-1015 (1988).

[114]   N. Takato, K. Jinguji, M. Yasu, H. Toba, and M. Kawachi, "Silica-based single-mode waveguides on silicon and their application to guided-wave optical interferometers," J. Lightwave Technol. **6**(6), 1003-1010 (1988).

[115] S. Suzuki, Y. Inoue, and T. Kominato, "High-density integrated 1×16 optical FDM multi/demultiplexer," in *Lasers and Electro-Optics Society Annual Meeting, 1994. LEOS '94 Conference Proceedings. IEEE* (1994), pp. 263-264 vol.262.

[116] N. Takato, T. Kominato, A. Sugita, K. Jinguji, H. Toba, and M. Kawachi, "Silica-based integrated optic Mach-Zehnder multi/demultiplexer family with channel spacing of 0.01-250 nm," IEEE J. Sel. Areas Commun. **8**(6), 1120-1127 (1990).

[117] K. Takiguchi, M. Oguma, T. Shibata, and H. Takahashi, "Demultiplexer for optical orthogonal frequency-division multiplexing using an optical fast-Fourier-transform circuit," Opt. Lett. **34**(12), 1828-1830 (2009).

[118] K. Takiguchi, M. Oguma, H. Takahashi, and A. Mori, "Integrated-optic eight-channel OFDM demultiplexer and its demonstration with 160 Gbit/s signal reception," Electron. Lett. **46**(8), 575-576 (2010).

[119] A. J. Lowery, D. Liang Bangyuan, and J. Armstrong, "Performance of optical OFDM in ultralong-haul WDM lightwave systems," J. Lightwave Technol. **25**(1), 131-138 (2007).

[120] A. Lowery and J. Armstrong, "Orthogonal-frequency-division multiplexing for dispersion compensation of long-haul optical systems," Opt. Express **14**(6), 2079-2084 (2006).

[121] J. Armstrong, "OFDM for optical communications," J. Lightwave Technol. **27**(3), 189-204 (2009).

[122] R. P. Giddings, X. Q. Jin, and J. M. Tang, "First experimental demonstration of 6Gb/s real-time optical OFDM transceivers incorporating channel estimation and variable power loading," Opt. Express **17**(22), 19727-19738 (2009).

[123] Q. Yang, S. Chen, Y. Ma, and W. Shieh, "Real-time reception of multi-gigabit coherent optical OFDM signals," Opt. Express **17**(10), 7985-7992 (2009).

[124] H. C. H. Mulvad, M. Galili, L. K. Oxenløwe, H. Hu, A. T. Clausen, J. B. Jensen, C. Peucheret, and P. Jeppesen, "Demonstration of 5.1 Tbit/s data capacity on a single-wavelength channel," Opt. Express **18**(2), 1438-1443 (2010).

[125] E. Yamada, A. Sano, H. Masuda, T. Kobayashi, E. Yoshida, Y. Miyamoto, Y. Hibino, K. Ishihara, Y. Takatori, K. Okada, K. Hagimoto, T. Yamada, and H. Yamazaki, "Novel no-guard-interval PDM CO-OFDM transmission in 4.1 Tb/s (50 × 88.8-Gb/s) DWDM link over 800 km SMF including 50-GHz spaced ROADM nodes," in *Optical Fiber Communication Conference* (2008), paper PDP8.

[126] R. Metcalfe, "Towards terabit ethernet," in *Optical Fiber Communication Conference* (2008), paper Keynote.

[127] A. Sano, H. Masuda, E. Yoshida, T. Kobayashi, E. Yamada, Y. Miyamoto, F. Inuzuka, Y. Hibino, Y. Takatori, K. Hagimoto, T. Yamada, and Y. Sakamaki, "30 × 100-Gb/s all-optical OFDM transmission over 1300 km SMF with 10 ROADM nodes," in *European Conference on Optical Communication* (2007), paper PDP1.7.

[128] G. Cincotti, "Fiber wavelet filters [and planar waveguide couplers for full-wavelength demultiplexers]," IEEE J. Quantum Electron. **38**(10), 1420-1427 (2002).

[129] Y.-K. Huang, D. Qian, R. E. Saperstein, P. N. Ji, N. Cvijetic, L. Xu, and T. Wang, "Dual-polarization 2x2 IFFT/FFT optical signal processing for 100-Gb/s QPSK-PDM all-optical OFDM," in *Optical Fiber Communication Conference* (2009), paper OTuM4.

[130] K. Takiguchi, M. Oguma, T. Shibata, and H. Takahashi, "Optical OFDM demultiplexer using silica PLC based optical FFT circuit," in *Optical Fiber Communication Conference* (2009), paper OWO3.

[131] D. Tse and P. Viswanath, *Fundamentals of wireless communication* (Cambridge University Press, 2005).

[132] R. Hassun, M. Flaherty, R. Matreci, and M. Taylor, "Effective evaluation of link quality using error vector magnitude techniques," in *Wireless Communications Conference* (1997), pp. 89-94.

[133] R. A. Shafik, S. Rahman, and A. H. M. R. Islam, "On the extended relationships among EVM, BER and SNR as performance metrics," in *International Conference on Electrical and Computer Engineering (ICECE)* (2006), pp. 408-411.

[134] T. Mizuochi, "Recent progress in forward error correction and its interplay with transmission impairments," IEEE J. Sel. Topics Quantum Electron. **12**(4), 544-554 (2006).

[135] S. Ramachandran, *Fiber based dispersion compensation* (Springer, 2007).

[136] P. Frascella, F. C. G. Gunning, S. K. Ibrahim, P. Gunning, and A. D. Ellis, "PMD tolerance of 288 Gbit/s Coherent WDM and transmission over unrepeatered 124 km of field-installed single mode optical fiber," Opt. Express **18**(13), 13908-13914 (2010).

[137] S. L. Jansen, "Multi-carrier approaches for next-generation transmission: Why, where and how?," in *Optical Fiber Communication Conference* (Optical Society of America, 2012), paper OTh1B.1.

[138] D. Kilper, "Energy efficient networks," in *Optical Fiber Communication Conference* (Optical Society of America, 2011), paper OWI5.

[139] K. Kitayama, S. Kocsis, T. C. Ralph, and G. Xiang, "Photonic networks beyond the next power-saving, security, and resilience," in *International Quantum Electronics Conference and Conference on Lasers and Electro-Optics Pacific Rim* (Optical Society of America, 2011), paper J1.

[140] C. Kachris, E. Giacoumidis, and I. Tomkos, "Energy-efficiency study of optical OFDM in data centers," in *Optical Fiber Communication Conference* (Optical Society of America, 2011), paper JWA087.

[141] B. Châtelain, C. Laperle, K. Roberts, X. Xu, M. Chagnon, A. Borowiec, F. Gagnon, J. Cartledge, and D. V. Plant, "Optimized pulse shaping for intra-channel nonlinearities mitigation in a 10 Gbaud dual-polarization 16-QAM system," in *Optical Fiber Communication Conference* (Optical Society of America, 2011), paper OWO5.

[142] C. Behrens, S. Makovejs, R. I. Killey, S. J. Savory, M. Chen, and P. Bayvel, "Pulse-shaping versus digital backpropagation in 224Gbit/s PDM-16QAM transmission," Opt. Express **19**(14), 12879-12884 (2011).

[143] J. Zhao and A. D. Ellis, "Electronic impairment mitigation in optically multiplexed multicarrier systems," J. Lightwave Technol. **29**(3), 278-290 (2011).

[144] S. Kilmurray, T. Fehenberger, P. Bayvel, and R. I. Killey, "Comparison of the nonlinear transmission performance of quasi-Nyquist WDM and reduced guard interval OFDM," Opt. Express **20**(4), 4198-4205 (2012).

[145] R. Schmogrow, D. Hillerkuss, S. Wolf, B. Bäuerle, M. Winter, P. Kleinow, B. Nebendahl, T. Dippon, P. C. Schindler, C. Koos, W. Freude, and J. Leuthold, "512QAM Nyquist sinc-pulse transmission at 54 Gbit/s in an optical bandwidth of 3 GHz," Opt. Express **20**(6), 6439-6447 (2012).

[146] B. Szafraniec, B. Nebendahl, and T. Marshall, "Polarization demultiplexing in Stokes space," Opt. Express **18**(17), 17928-17939 (2010).

[147] R. Schmogrow, B. Nebendahl, M. Winter, A. Josten, D. Hillerkuss, S. Koenig, J. Meyer, M. Dreschmann, M. Huebner, and C. Koos, "Corrections to "Error Vector

Magnitude as a Performance Measure for Advanced Modulation Formats"," IEEE Photon. Technol. Lett. **24**(23), 2198-2198 (2012).

[148] N. Lindenmann, G. Balthasar, D. Hillerkuss, R. Schmogrow, M. Jordan, J. Leuthold, W. Freude, and C. Koos, "Photonic wire bonds for terabit/s chip-to-chip interconnects," ArXiv e-prints, arXiv:1111.0651v1 (2011),

http://arxiv.org/abs/1111.0651v1.

[149] N. Lindenmann, G. Balthasar, D. Hillerkuss, R. Schmogrow, M. Jordan, J. Leuthold, W. Freude, and C. Koos, "Photonic wire bonding: a novel concept for chip-scale interconnects," Opt. Express **20**(16), 17667-17677 (2012).

[150] R. Bonk, P. Vorreau, D. Hillerkuss, W. Freude, G. Zarris, D. Simeonidou, F. Parmigiani, P. Petropoulos, R. Weerasuriya, S. Ibrahim, A. D. Ellis, D. Klonidis, I. Tomkos, and J. Leuthold, "An all-optical grooming switch for interconnecting access and metro ring networks [Invited]," J. Opt. Commun. Netw. **3**(3), 206-214 (2011).

[151] D. Hillerkuss, "Implementation of an optical clock recovery and OTDM-to-WDM conversion," Diploma Thesis No. 786, *Institute of High-Frequency and Quantum Electronics (IHQ)* (University of Karlsruhe (TH), Karlsruhe, 2008).

[152] P. Vorreau, S. Sygletos, F. Parmigiani, D. Hillerkuss, R. Bonk, P. Petropoulos, D. J. Richardson, G. Zarris, D. Simeonidou, D. Klonidis, I. Tomkos, R. Weerasuriya, S. Ibrahim, A. D. Ellis, D. Cotter, R. Morais, P. Monteiro, S. Ben-Ezra, S. Tsadka, W. Freude, and J. Leuthold, "Optical grooming switch with regenerative functionality for transparent interconnection of networks," Opt. Express **17**(17), 15173-15185 (2009).

[153] G. Zarris, E. Hugues-Salas, N. A. Gonzalez, R. Weerasuriya, F. Parmigiani, D. Hillerkuss, P. Vorreau, M. Spyropoulou, S. K. Ibrahim, A. D. Ellis, R. Morais, P. Monteiro, P. Petropoulos, D. J. Richardson, I. Tomkos, J. Leuthold, and D. Simeonidou, "Field experiments with a grooming switch for OTDM meshed networking," J. Lightwave Technol. **28**(4), 316-327 (2010).

[154] G. Zarris, F. Parmigiani, E. Hugues-Salas, R. Weerasuriya, D. Hillerkuss, N. A. Gonzalez, M. Spyropoulou, P. Vorreau, R. Morais, S. K. Ibrahim, D. Klonidis, P. Petropoulos, A. D. Ellis, P. Monteiro, A. Tzanakaki, D. Richardson, I. Tomkos, R. Bonk, W. Freude, J. Leuthold, and D. Simeonidou, "Field trial of WDM-OTDM transmultiplexing employing photonic switch fabric-based buffer-less bit-interleaved data grooming and all-optical regeneration," in *Optical Fiber Communication Conference* (2009), paper PDPC10.

[155] G. Zarris, P. Vorreau, D. Hillerkuss, S. K. Ibrahim, R. Weerasuriya, A. D. Ellis, J. Leuthold, and D. Simeonidou, "WDM-to-OTDM traffic grooming by means of asynchronous retiming," in *Optical Fiber Communication Conference* (2009), paper OThJ6.

[156] P. Vorreau, D. Hillerkuss, F. Parmigiani, S. Sygletos, R. Bonk, P. Petropoulos, D. Richardson, G. Zarris, D. Simeonidou, D. Klonidis, I. Tomkos, R. Weerasuriya, S. Ibrahim, A. Ellis, R. Morais, P. Monteiro, S. Ben Ezra, S. Tsadka, W. Freude, and J. Leuthold, "2R/3R optical grooming switch with time-slot interchange," in *European Conference on Optical Communication* (2008), paper PDP Th.3.F.4. .

[157] L. Alloatti, D. Korn, D. Hillerkuss, T. Vallaitis, J. Li, R. Bonk, R. Palmer, T. Schellinger, A. Barklund, R. Dinu, J. Wieland, M. Fournier, J. Fedeli, P. Dumon, R. Baets, C. Koos, W. Freude, and J. Leuthold, "40 Gbit/s silicon-organic hybrid (SOH) phase modulator," in *European Conference on Optical Communication* (2010), paper Tu.5.C.4.

[158] L. Alloatti, D. Korn, D. Hillerkuss, T. Vallaitis, J. Li, R. Bonk, R. Palmer, T. Schellinger, C. Koos, W. Freude, J. Leuthold, M. Fournier, J. Fedeli, A. Barklund, R.

Dinu, J. Wieland, W. Bogaerts, P. Dumon, and R. Baets, "Silicon high-speed electro-optic modulator," in *IEEE International Conference on Group IV Photonics* (2010), pp. 195-197.

[159]   L. Alloatti, D. Korn, R. Palmer, D. Hillerkuss, J. Li, A. Barklund, R. Dinu, J. Wieland, M. Fournier, J. Fedeli, H. Yu, W. Bogaerts, P. Dumon, R. Baets, C. Koos, W. Freude, and J. Leuthold, "42.7 Gbit/s electro-optic modulator in silicon technology," Opt. Express **19**(12), 11841-11851 (2011).

[160]   W. Freude, L. Alloatti, A. Melikyan, R. Palmer, D. Korn, N. Lindenmann, T. Vallaitis, D. Hillerkuss, J. Li, A. Barklund, R. Dinu, J. Wieland, M. Fournier, J.-M. Fedeli, S. Walheim, P. Leufke, S. Ulrich, J. Ye, P. Vincze, H. Hahn, H. Yu, W. Bogaerts, P. Dumon, R. Baets, B. Breiten, F. Diederich, M. T. Beels, I. Biaggio, T. Schimmel, C. Koos, and J. Leuthold, "Nonlinear optics on the silicon platform," in *Optical Fiber Communication Conference* (Optical Society of America, 2012), paper OTh3H.6.

[161]   W. Freude, L. Alloatti, T. Vallaitis, D. Korn, D. Hillerkuss, R. Bonk, R. Palmer, J. Li, T. Schellinger, M. Fournier, J. Fedeli, W. Bogaerts, P. Dumon, R. Baets, A. Barklund, R. Dinu, J. Wieland, M. L. Scimeca, I. Biaggio, B. Breiten, F. Diederich, C. Koss, and J. Leuthold, "High-speed signal processing with silicon-organic hybrid devices," in *Journal of the European Optical Society: rapid publications* (2010).

[162]   T. Vallaitis, R. Bonk, J. Guetlein, D. Hillerkuss, J. Li, W. Freude, J. Leuthold, C. Koos, M. L. Scimeca, I. Biaggio, F. Diederich, B. Breiten, P. Dumon, and R. Baets, "All-optical wavelength conversion of 56 Gbit/s NRZ-DQPSK signals in silicon-organic hybrid strip waveguides," in *Optical Fiber Communication Conference* (2010), paper OTuN1.

[163]   T. Vallaitis, D. Hillerkuss, J. S. Li, R. Bonk, N. Lindenmann, P. Dumon, R. Baets, M. L. Scimeca, I. Biaggio, F. Diederich, C. Koos, W. Freude, and J. Leuthold, "All-optical wavelength conversion using cross-phase modulation at 42.7 Gbit/s in silicon-organic hybrid (SOH) waveguides," in *Photonics in Switching* (2009), pp. 1-2.

[164]   H. Yu, W. Bogaerts, K. Komorowska, R. Baets, D. Korn, L. Alloatti, D. Hillerkuss, C. Koos, W. Freude, J. Leuthold, J. Campenhout, P. Verheyen, J. Wouters, M. Moelants, and P. Absil, "Doping geometries for 40G carrier-depletion-based Silicon optical modulators," in *Optical Fiber Communication Conference* (Optical Society of America, 2012), paper OW4F.4.

[165]   W. Freude, R. Schmogrow, B. Nebendahl, M. Winter, A. Josten, D. Hillerkuss, S. Koenig, J. Meyer, M. Dreschmann, M. Huebner, C. Koos, J. Becker, and J. Leuthold, "Quality metrics for optical signals: eye diagram, Q-factor, OSNR, EVM and BER," in *International Conference on Transparent Optical Networks* (2012), paper Mo.B1.5.

[166]   M. Jordan, "Comb generation for next generation optical transmission systems," Diplomarbeit No. 846, *Institute of Photonics and Quantum Electronics (IPQ)* (Karlsruhe Institute of Technology (KIT), Karlsruhe, 2012).

[167]   I. S. Gradstein and I. M. Ryshik, *Tables of series, products, and integrals* (Harri Deutsch, Thun, 1981).

[168]   M. Abramowitz and I. A. Stegun, "Exponential integral and related functions – Equation 5.2.1 on page 231," in *Handbook of Mathematical Functions*(Dover, New York, 1972).

[169]   S. H. Rowe, "Efficiency enhancement of light modulators," Appl. Opt. **9**(5), 1222-1222 (1970).

[170]   L. Zehnder, "Ein neuer Interferenzrefraktor," Zeitschrift für Instrumentenkunde **11**(275-285 (1891).

[171]   L. Mach, "Ueber einen Interferenzrefraktor," Zeitschrift für Instrumentenkunde **12**(89-93 (1892).

[172]   J. Leuthold, "Optical transmitters and receivers - Part 2: Transmitters and modulation formats," (Karlsruhe, 2011).

[173]   D. A. B. Miller, D. S. Chemla, T. C. Damen, A. C. Gossard, W. Wiegmann, T. H. Wood, and C. A. Burrus, "Band-edge electroabsorption in quantum qell structures - the quantum-confined Stark-effect," Phys. Rev. Lett. **53**(22), 2173-2176 (1984).

[174]   R. K. Willardson, W. T. Tsang, and A. C. Beer, *Semiconductors and semimetals: lightwave communications technology. photodetectors* (Academic Press, 1985).

[175]   M. Jeruchim, "Techniques for estimating the bit error rate in the simulation of digital communication systems," IEEE J. Sel. Areas Commun. **2**(1), 153-170 (1984).

[176]   M. Bischoff and G. Grau, "Optical CPFSK receiver with BER <10^-9: Monte Carlo simulation and experiment," Electron. Lett. **29**(10), 899-900 (1993).

[177]   M. Bischoff, "Simulation kohärent-optischer Übertragungssysteme am Beispiel eines CPFSK-Heterodynsystems," PhD Thesis No. ISBN: 3-18-328110-4, (University of Karlsruhe, 1993).

[178]   V. Ribeiro, L. Costa, M. Lima, and A. L. J. Teixeira, "Optical performance monitoring using the novel parametric asynchronous eye diagram," Opt. Express **20**(9), 9851-9861 (2012).

[179]   A. Papoulis and S. U. Pillai, *Probability, random variables, and stochastic processes* (McGraw-Hill, 2002).

[180]   S. D. Personick, "Receiver design for digital fiber optic communication systems, I," Bell System Technical Journal **52**(6), 843-874 (1973).

[181]   T. Kremp and W. Freude, "Fast split-step wavelet collocation method for WDM system parameter optimization," J. Lightwave Technol. **23**(3), 1491-1502 (2005).

[182]   D. Marcuse, "Derivation of analytical expressions for the bit-error probability in lightwave systems with optical amplifiers," J. Lightwave Technol. **8**(12), 1816-1823 (1990).

[183]   D. Marcuse, "Calculation of bit-error probability for a lightwave system with optical amplifiers and post-detection Gaussian noise," J. Lightwave Technol. **9**(4), 505-513 (1991).

[184]   W. Freude, "Positive definite and unimodal Gauss-Hermite expansion of probability density function by its moments," AEÜ **45**(89-95 (1991).

[185]   M. Dlubek, A. Phillips, and E. Larkins, "Optical signal quality metric based on statistical moments and Laguerre expansion," Optical and Quantum Electronics **40**(8), 561-575 (2008).

[186]   H. A. Mahmoud and H. Arslan, "Error vector magnitude to SNR conversion for nondata-aided receivers," Wireless Communications, IEEE Transactions on **8**(5), 2694-2704 (2009).

[187]   L. Lo Presti and M. Mondin, "Design of optimal FIR raised-cosine filters," Electron. Lett. **25**(7), 467-468 (1989).

[188]   E. Sun, B. Tian, Y. Wang, and K. Yi, "Quasi-orthogonal time division multiplexing and its applications under Rayleigh fading channels," in *Advanced Information Networking and Applications Workshops (AINAW)* (2007), pp. 172-176.

[189] P. A. Milder, R. Bouziane, R. Koutsoyannis, C. R. Berger, Y. Benlachtar, R. I. Killey, M. Glick, and J. C. Hoe, "Design and simulation of 25 Gb/s optical OFDM transceiver ASICs," Opt. Express **19**(26), B337-B342 (2011).

[190] D. Rafique and A. D. Ellis, "Nonlinear penalties in long-haul optical networks employing dynamic transponders," Opt. Express **19**(10), 9044-9049 (2011).

[191] Xilinx, "Virtex-5 FPGA RocketIO GTX Transceiver User Guide - UG198," (Xilinx, Inc.).

[192] Y. Chiu, B. Jalali, S. Garner, and W. Steier, "Broad-band electronic linearizer for externally modulated analog fiber-optic links," IEEE Photon. Technol. Lett. **11**(1), 48-50 (1999).

Appendix C: Glossary

C.1 List of Symbols

Greek Symbols

α	Probability that a measured BER lies within a certain confidence interval ε
β_2	Local group velocity dispersion GVD coefficient
γ	Electrical signal-to-noise power ratio
γ_{OA}	Shot-noise limited electrical signal-to-noise ratio
ε	Confidence Interval
η	Quantum efficiency of a photodiode
ϑ	Phase of the optical carrier, see eq. (2.1)
λ	Optical wavelength
λ_O	Wavelength of the optical carrier in a receiver
σ_{err}	Root mean square if various error vectors E_{err}
σ_1	Standard deviation of the PDF of the "1" level in an eye diagram
σ_0	Standard deviation of the PDF of the "0" level in an eye diagram
τ	Delay of a delay interferometer
τ_{GI}	Duration of a guard interval in case of OFDM
τ_{BER}	BER measurement time for given confidence
Φ	Initial carrier phase of the modulator and overall delay phase delay
$\varphi, \varphi_m, \varphi_n$	Phase shift in a Mach-Zehnder modulator, delay interferometer, or FFT structure m, $n = 1, 2, 3, \ldots$
φ_1, φ_2	Phases of the phase modulators in a Mach-Zehnder modulator
φ_I, φ_Q	Phase difference in the in-phase (I) or quadrature phase (Q) Mach-Zehnder modulator in an IQ-modulator
$\varphi_{I,1}, \varphi_{I,2}$	Phases of the phase modulators in the in-phase Mach-Zehnder modulator in an IQ-modulator

$\varphi_{Q,1}, \varphi_{Q,2}$	Phases of the phase modulators in the quadrature phase Mach-Zehnder modulator in an IQ-modulator
$\Delta\omega$	Angular carrier spacing of an OFDM signal
ω_F	Angular center frequency of a Gaussian filter

Latin Symbols

$\mathbf{A}$	Phasor of a complex signal, sometimes also referred to as complex field vector (2.5)
A	Amplitude of an optical carrier or of a signal
B	Bandwidth of a channel
B^{Nyq}	Bandwidth of a Nyquist signal
BER	Bit error probability or bit error ratio
$\mathrm{BER}_{\mathrm{op}}$	BER at the optimum threshold
BR	Bitrate of a channel
B_E	Electrical filter bandwidth in a direct detection receiver
B_{OFDM}	Bandwidth of an OFDM signal
B_{O}	Optical filter bandwidth in a receiver
B_{ref}	Reference bandwidth for $\mathrm{OSNR}_{\mathrm{ref}}$
B_2	Accumulated group velocity dispersion (GVD)
b	Number of bits transmitted per symbol
b_n	n^{th} bit of a PRBS sequence
C	Information capacity of a data transmission (measured in bit/s)
C_{DI}	Complexity of the FFT with delay interferometers
C_{std}	Complexity of the direct FFT implementation
CP	Amount of cyclic prefix with respect to the effective OFDM symbol duration $\mathrm{CP} = \tau_c / \left(T_s + \tau_c\right)$
c	Speed of light in vacuum
c_{ik}	Complex modulation coefficients of an OFDM signal or an aggregated Nyquist WDM frequency band
c'_{ki}	Received complex modulation coefficients of an OFDM signal – ideally equal to c_{ik}

E	Electric field strength in front of photo detector in a direct detection receiver
$E(t)$	Electrical field of the optical carrier, see eq. (2.1)
E_{err}	Error vector of a received phasor
$E_{err,i}$	Error vector of the i-th symbol
E_m	Size of the even DFT when performing the FFT
E_r	Received phasor of a signal
$E_{r,i}$	Received phasor of the i-th symbol
E_t	Transmitted phasor of a signal
$E_{t,i}$	Transmitted phasor of the i-th symbol
$\left|E_{t,a}\right|$	Average field resulting from summing the powers of all M possible symbols
$\left|E_{t,m}\right|$	Maximum field in the constellation
F_s	Nyquist bandwidth of a signal, spectral width of a Nyquist signal, typically equal to the symbol rate F_s^{Nyq}
F_s^{Nyq}	Symbol rate of a Nyquist signal
F_s^{OFDM}	Symbol rate of an OFDM signal
e	Elementary charge
EVM	Error vector magnitude
EVM_a	EVM normalized to the average field $\left|E_{t,a}\right|$
EVM_m	EVM normalized to the maximum field in the constellation
EVM_{tot}	Total EVM of a super-channel
F	Noise figure
F_s	Sampling frequency
f	Frequency
f_{RF}	RF-carrier frequency of an OFDM signal
f_c	Carrier frequency
f_k	Carrier frequencies (also with $k = 1, 2, 3$)
f_L	Frequency of the incident light at a photodiode
$f_{\max}$	Maximum frequency in a signal without aliasing

f_O	Optical carrier frequency
f_{sa}	Maximum frequency in a signal without aliasing
Δf	Carrier spacing of a signal
G	Power gain of an optical amplifier
H	Transfer function of a modulator
$H(f)$	Channel transfer function
$H_m(\omega)$	Transfer function of the m-th FFT output in the frequency domain
h	Planck's constant
h_r	Filter coefficients of an FIR filter
$h(t)$	Received voltage impulse, e. g. raised cosine
I	In-phase component of a signal
$I,\mathbf{I}$	In-phase axis in a constellation or time diagram
i	Generated electrical current in a photo detector
K	Whole number
k,l	Subcarrier indices
l	In A.6.6: length of the PRBS sequence 2^N-1+1
L	Transmission fiber length
M	Number of constellation points of a given modulation format (e.g. M-QAM or M-PSK)
m	Frequency index of a subcarrier of an OFDM symbol at the FFT input or IFFT output In A.6.6: Number of inputs of an m:1 multiplexer
N	Number of time or frequency samples processed by the FFT or IFFT, Number of OFDM carriers, Number of carriers in a Nyquist WDM system In A.6.6 length of the sections of the PRBS sequence to be multiplexed.
N_S	Number of analyzed symbols for computation of the error vector magnitude
N_R	Number of bits per sample in a DAC

n	Time index of a sample of an OFDM symbol the FFT output or IFFT input In A.6.6 $n \in \mathbb{N}$
n_{sp}	Spontaneous emission factor
O_m	Size of the odd DFT when performing the FFT
OSNR	Optical signal-to-noise ratio. Signal power with respect to the polarized ASE power in the actual optical receiver bandwidth B_O
$OSNR_{ref}$	Optical signal-to-noise ratio with fixed reference bandwidth. Signal power related to the unpolarized ASE power in a fixed reference bandwidth
$\overline{P}$	Average power of a Nyquist signal with sinc-pulses
$P_{ASE}^{x,\,pol}$	Polarized noise power within bandwidth B_x
$P_{ASE}^{O,\,pol}$	Polarized ASE power in the actual optical receiver bandwidth B_O
P_e	Optical Power at the input of a photodiode
$P_{S,\,a}$	Average signal power in a receiver
$P_S^{(1)}$	Signal power of a typical voltage impulse
P_{max}	Maximum peak power of a Nyquist signal with sinc-pulses
P_N	Average noise power in a receiver
P_O	Optical receiver input power
$PAPR_{Nyquist}$	Peak-to-average power ratio of a Nyquist signal
$PAPR_{OFDM}$	Peak-to-average power ratio of an OFDM signal
p	Number of stages in an FFT of the order $N = 2^p$ or index of an FFT stage
$p(1t)$	Probability of transmitting a "1"
$p(0t)$	Probability of transmitting a "0"
Q	Quadrature phase component of a signal
Q_F	Quality factor of a signal as defined in eq. (8.37)
$Q,\mathbf{Q}$	Quadrature axis in constellation or time diagram
q	Oversampling factor
R	Symbol rate of a signal
R_O	Order of an FIR-filter

r	Exponent in BER $= 10^{-r}$
$r\left(t+iT_s\right)$	Received OFDM signal
$r^{(i)}(t)$	Received OFDM symbol
$S_{21}(f)$	Optical filter transfer function
SE	Spectral efficiency of a signal, see eq. (1.1)
$\mathrm{SE}_{\mathrm{Nyquist}}$	Spectral efficiency of a Nyquist signal
$\mathrm{SE}_{\mathrm{OFDM}}$	Spectral efficiency of an OFDM signal
T	Symbol duration of a signal
T_s	OFDM: Temporal width of symbol, defined as $T_s = 1/\Delta f$, Nyquist: temporal subcarrier spacing
T_{sa}	Temporal distance of sampling points of a discrete signal.
t	Time
t_k	Temporal position of the Nyquist pulse of symbol k
u	Voltage after detection in a direct detection receiver
$u_{\mathrm{th,op}}$	Optimum decision threshold for a minimum $\mathrm{BER}_{\mathrm{op}}$
u_{th}	Threshold voltage
u_1	Average voltage of the "1" level in an eye diagram
u_0	Average voltage of the "0" level in an eye diagram
$w_1(u)$	Probability density function for the "1" level in an eye diagram
$w_0(u)$	Probability density function for the "0" level in an eye diagram
$X(f)$	Spectrum of the OFDM signal
$X^{(i)}(f)$	Spectrum of the i-th symbol of an OFDM signal
$X_k(f)$	Spectrum of one OFDM subcarrier within an OFDM symbol
X_m	Data encoded on the m-th subcarrier of an OFDM symbol, also m-th frequency sample of a signal in the frequency domain
$X_m(t)$	Continuous output signal of an FFT structure
$\hat{X}_m(\omega)$	Fourier transform of the continuous output signal of an FFT.
$x(t)$	OFDM signal, *infinite* sequence of *temporal* symbols $x^{(i)}(t)$
$\hat{x}(\omega)$	Fourier transform of the OFDM signal $x(t)$.
$x^{(i)}(t)$	*Temporal* symbols superscripted with i of an OFDM signal $x(t)$

x_n	n-th time sample of an OFDM symbol, also m-th sample of a signal in the time domain
$Y(f)$	Spectrum of a Nyquist WDM band with N pulse shaped channels
$Y^{(i)}(f)$	The ith spectral Nyquist symbol
$Y_k(f)$	Spectrum of one Nyquist pulse
$y(t)$	Nyquist signal
$y^{(i)}$	Train of modulated Nyquist-pulses
y_k	Single Nyquist pulse
z	Variable

C.2 Acronyms

16QAM	16ary quadrature amplitude modulation
64QAM	64ary quadrature amplitude modulation
ADC	Analog-to-digital converter
ASE	Amplified spontaneous emissions
ASK	Amplitude shift keying
AWG	Arbitrary waveform generator
AWGN	Additive white Gaussian noise
B2B	Back-to-back, characterization without transmission link
BER	Bit error ratio
BERT	Bit error tester
BPSK	Binary phase shift keying
C	Optical coupler
CD	Chromatic dispersion
CDB	Clock domain bridge
CLK	Sampling clock frequency
Co-WDM	Coherent wavelength division multiplexing
CP	Cyclic prefix or cyclic postfix
DAC	Digital-to-analog converter
DCA	Digital communications analyzer - Agilent Oscilloscope

DCF	Dispersion compensating fiber
DFT	Discrete Fourier transform
DI	Delay interferometer
DIL	Dis-interleaver
DMX	Demultiplexer
DP-16QAM	Dual polarization 16ary quadrature amplitude modulation
DP-BPSK	Dual polarization binary phase shift keying
DP-QPSK	Dual polarization quadrature phase shift keying
DPSK	Differential phase shift keying
DQPSK	Differential quadrature phase shift keying
DRP	Dynamic reconfiguration port
DSP	Digital signal processing
EA	Electronic Amplifier
EAM	Electro-absorption modulator
ECL	External cavity laser
EDFA	Erbium doped fiber amplifier
EVM	Error vector magnitude
FDM	Frequency division multiplexing
FEC	Forward error correction
FFT	Fast Fourier transform
FIFO	First-in first-out shift register
FIR	Finite impulse response filter
FPGA	Field programmable gate array
FSK	Frequency shift keying
FSR	Free spectral range of a filter
GVD	Group velocity dispersion
HNLF	Highly nonlinear fiber
IDFT	Discrete Fourier transform
IFFT	Inverse fast Fourier transform
IL	Interleaver

IOFFT	Inverse optical fast Fourier transform
IQ-modulator	Modulator for complex modulation
ISI	Intersymbol interference
ITO	Indium tin oxide
ITU	International Telecommunication Union
ITU-T	Telecommunication standardization sector of ITU
LCoS	Liquid crystal on silicon
LiNbO3	Lithium Niobate
LO	Local oscillator
MCM	Multi carrier modulation
MGT	Multi gigabit transceivers
MLL	Mode-locked laser
MUX	Multiplexer
MZM	Mach-Zehnder modulator
NLE	Nonlinear element
NRZ	Non-return to zero
Nyquist WDM	Nyquist wavelength division multiplexing
OA	Optical pre-amplifier
OFDM	Orthogonal frequency division multiplexing
OFFT	Optical fast Fourier transform
OMA	Agilent optical modulation analyzer N4391A
OOK	On-off-keying
OSA	Optical spectrum analyzer
OSNR	Optical signal to noise ratio
OTDM	Optical time division multiplexing
PAM4	Pulse amplitude modulation with 4 signal levels
PAPR	Peak-to-average power ratio
pdf	Probability density function
PDM	Polarization division multiplexing
PLB	Processor local bus

PMD	Polarization mode dispersion
PolMUX	Polarization multiplexing
Pol-SK	Polarization shift keying
PRBS	Pseudo random bit sequence
PS, P/S	Parallel-to-serial conversion
PSK	Phase shift keying
QAM	Quadrature amplitude modulation
QPSK	Quadrature phase shift keying
RF	Radio frequency
RFS	Recirculating frequency shifter
RX	Receiver
SE	Spectral efficiency
SNR	Signal to noise ratio
SOA	Semiconductor optical amplifier
SP, S/P	Serial-to-parallel conversion
SP-BPSK	Single polarization binary phase shift keying
SSMF, SMF-28	Standard single mode fiber
STFT	Short-time Fourier Transform
TDM	Time division multiplexing
TX	Transmitter
UART	Universal asynchronous receiver transmitter
VHDL	VHSIC hardware description language
VHSIC	Very high speed integrated circuit
WDM	Wavelength division multiplexing
WGR	Waveguide grating router
WS	Waveshaper
WSS	Wavelength selective switch
XOR	Exclusive OR logic gate

Acknowledgements

Die vorliegende Dissertation entstand während meiner Tätigkeit am Institut für Photonik und Quantenelektronik (IPQ) des Karlsruher Instituts für Technologie (KIT). Sie war zum Teil eingebunden in die Europäischen Forschungsprojekte ACCORDANCE, TRIUMPH und Euro-Fos, das vom Bundesministerium für Bildung und Forschung geförderte BMBF-Projekt CONDOR sowie in die Karlsruhe School of Optics and Photonics (KSOP), gefördert von der Deutschen Forschungsgemeinschaft (DFG). Am Ende meiner Arbeit möchte ich all denjenigen Personen meinen herzlichen Dank aussprechen, die im Verlauf der letzten Jahre zum Gelingen des vorliegenden Manuskriptes beigetragen haben.

An erster Stelle danke ich meinem Doktorvater Prof. sc. nat. Dr. J. Leuthold für das entgegengebrachte Vertrauen, die massive Unterstützung meiner Arbeit, das große und stetige Interesse an meiner Arbeit, die zahlreichen Anregungen und seine innovativen Ideen. Sein fachlicher Überblick sowie sein Blick für das große Ganze trugen essentiell zum Gelingen der Arbeit bei.

Herrn Prof. Dr. Dr. h.c. W. Freude, meinem zweiten Betreuer, danke ich für die akribische Durchsicht meiner vielen Manuskripte und die immer lehrreichen fachlichen Diskussionen. Seine vorbildliche Genauigkeit und der Drang zur Perfektion ist eine unvergessliche Schule fürs weitere Leben. Seine große fachliche Kompetenz und seine Erfahrung gewährleistete immer ein hohes wissenschaftliches Niveau.

Herrn Prof. Dr.-Ing. Michael Hübner danke ich für die freundliche Übernahme des Korreferats und die stete Unterstützung meiner Arbeit.

Prof. Dr. Andrew Ellis, dem Betreuer meines Industriepraktikums und meiner Studienarbeit danke ich für sein Engagement und seine Unterstützung während meiner Zeit in Irland. Die dort erlernten Fähigkeiten wie die Planung und die Durchführung von Experimenten und beim Arbeiten im Labor haben mir den Start in meine Doktorarbeit stark erleichtert.

Dr. Philipp Vorreau, dem Betreuer meiner Diplomarbeit und Kollegen danke ich für seine Unterstützung und die hervorragende Einarbeitung in das Labor des Institutes.

Meinen Kollegen Dr. Thomas Vallaitis, Dr. René Bonk und René Schmogrow danke ich für die langjährige erfolgreiche Zusammenarbeit. Die vielen gemeinsamen Messungen und unzähligen Diskussionen über physikalische wie elektro-technische Zusammenhänge bildeten ein sehr wichtiges Fundament zur Durchführung und dem Gelingen der Arbeit. Das gegenseitige Unterstützung bei Experimenten und das Vertrauen im Umgang mit Ideen und Veröffentlichungen machten ein erfolgreiches Arbeiten möglich.

Prof. Dr. Michael Huebner, Michael Dreschmann, Joachim Meyer und Prof. Dr. Jürgen Becker vom ITIV danke ich für ihre stete Unterstützung unserer Arbeit.

Ich möchte auch den vielen Partnern aus Industrie und Forschung danken, die mich in meinen Experimenten mit Rat, Tat und Equipment unterstützt haben: Shalva Ben-Ezra von Finisar, Israel, Bernd Nebendahl, Thomas Stefany, Thomas Dippon, Emmerich Müller und Matthias Kohler von Agilent Technologies, Francesca Parmigiani, Periklis Petropoulos, Xin Yang und David Richardson vom Optoelectronics Research Center in Southampton, Bojan

Resan, Andreas Öhler und Kurt Weingarten von Time Bandwidth Products, Tobias Ellermeyer, Joachim Lutz, Lars Altenhain, und Matthias Tom Frey von Micram, Michael Moeller von der Saarland Universität in Saarbrücken, Marc Eichhorn vom Deutsch-Französischen Forschungsinstitut in Saint-Louis, Sander Jansen von Nokia Siemens Networks und Beril Inan von der Technischen Universität München, sowie Erel Granot vom Ariel Universitätszentrum von Samaria.

Für die gute Zusammenarbeit bzw. Unterstützung beim Aufbau der Demonstratoren für das EU-Projekt Triumph danke ich Phlipp Vorreau, F. Parmigiani, René Bonk, George Zarris, E-milio Hugues-Salas, Maria Spyropoulou, Ruwan Weerasuriya, Shalva Ben-Ezra, Dimitrios Klonidis, Selwan Ibrahim, Periklis Petropoulos, Andrew Ellis, David Richardson, Dimitra Simeonidou, Ioannis Tomkos, Rui Morais und Paolo Monteiro.

Für die sorgfältige Durchsicht des vorliegenden Manuskriptes bedanke ich mich recht herzlich bei René Bonk, Philipp Schindler und René Schmogrow.

Meinen Bürokollegen René Schmogrow, Dr. Philipp Vorreau, Andrej Marculescu, Dr. Marcus Winter und Dr. Arvind Mishra danke ich für die freundliche und herzliche Zusammenarbeit über die Jahre, in denen man viel miteinander Durchleben konnte.

Ein großer Dank geht an „meine" zahlreichen motivierten und sehr guten Studenten, die direkt zum Gelingen dieser Arbeit beigetragen haben: Felix Frey, Meinert Jordan, Philipp Kleinow, Gregor Ronniger, Thomas Schellinger, René Schmogrow, Geirfinnur Smári Sigurðsson, Michael Teschke und Stefan Wolf. Nicht vergessen sind auch die weiteren Studenten, die ich während meiner Zeit am IPQ betreut habe und die mit ihrem großen Interesse stets dazu beigetragen haben meinen Horizont zu erweitern: Hammam Shakhtour, Djorn Karnick, Alejandro Sánchez, Lu Zhang, Yinfei Lu, Muhammad Rodlin Billah, Lin Lily, Hanna Almeye Bayu und Ho Hoai Duc Nguyen.

Ich möchte auch den Mitarbeitern des IPQs danken, die mich in den vergangenen Jahren mit einer Vielzahl von Arbeiten unterstützt haben. Bei Frau Bernadette Lehmann, Frau Ilse Kober, Frau Angelika Olbrich und Frau Andrea Riemensperger bedanke ich mich für die administrative Unterstützung. Vor allem Bernadette danke ich für die vielen leckeren Kekse und Gummibärchen. Ein besonderer Dank geht auch an Martin Winkeler und Johann Hartwig Hausschild, die mich in Sachen Computer, Elektronik und Bauteile immer unterstützt haben. Ihre Fähigkeiten, schnell Probleme zu lösen, haben mich oft vor großen Verzögerungen bewahrt. Zu Dank bin ich auch Oswald Speck aus dem Packaging-Labor für die sehr zuverlässige Organisation von Laborverbrauchsmitteln verpflichtet. Auch Herrn Bürger, Herrn Hirsch und Herrn Höhne, sowie allen Auszubildenden aus der mechanischen Werkstatt sage ich ein großes Dankeschön für die vielen Arbeiten und die immer tatkräftige Unterstützung. Sebastian Struck danke ich für die Unterstützung im Bereich Computer und Administration.

Bei allen Kollegen des IPQ bedanke ich mich recht herzlich für die gute Zusammenarbeit, Kooperation, den Spaß und die Erfahrungen auch außerhalb des IPQ: René Bonk, Swen König, Moritz Röger, Dr. Thomas Vallaitis, Prof. Dr. Christian Koos, Nicole Lindenmann, René Schmogrow, Philipp Schindler, Alexandra Ludwig, Dr. Marcus Winter, Dr. Arvind Mishra, Dr. Sean O'Duill, Dr. Philipp Vorreau, Dr. Stelios Sygletos, Luca Alloatti, Dietmar Korn,

Jinshi Li, Christos Klamouris, Argishti Melikyan, Sascha Mühlbrandt, Robert Palmer, Jörg Pfeifle, Simon Schneider, Claudius Weimann, Kai Worms und Frans Wegh.

Ganz besonders möchte ich Nicole, René (B. & S.), Tom, Swen, Philipp (V. & S.), Claudius, Micha, Moritz, Andrej und Bernadette für die Ablenkung und Unterstützung außerhalb des Institutes danken.

Vielen Dank auch an Beate, Andreas, Johannes, Benedikt, Simon, Lukas, Jakob, Ulrike, Clara und Maria.

List of Publications

[H1] A. D. Ellis, D. Cotter, S. Ibrahim, R. Weerasuriya, C. W. Chow, J. Leuthold, W. Freude, S. Sygletos, P. Vorreau, R. Bonk, D. Hillerkuss, I. Tomkos, A. Tzanakaki, C. Kouloumentas, D. J. Richardson, P. Petropoulos, F. Parmigiani, G. Zarris, and D. Simeonidou, "Optical interconnection of core and metro networks [Invited]," J. Opt. Netw. 7(11), 928-935 (2008).

[H2] D. Hillerkuss, "Implementation of an optical clock recovery and OTDM-to-WDM conversion," Diploma Thesis No. 786, *Institute of High-Frequency and Quantum Electronics (IHQ)* (University of Karlsruhe (TH), Karlsruhe, 2008).

[H3] D. Hillerkuss, A. D. Ellis, G. Zarris, D. Simeonidou, J. Leuthold, and D. Cotter, "40 Gbit/s asynchronous digital optical regenerator," Opt. Express **16**(23), 18889-18894 (2008).

[H4] S. K. Ibrahim, D. Hillerkuss, R. Weerasuriya, G. Zarris, D. Simeonidou, J. Leuthold, and A. D. Ellis, "Novel 42.65 Gbit/s dual gate asynchronous digital optical regenerator using a single MZM," in *European Conference on Optical Communication* (2008), paper Tu.4.D.3.

[H5] S. K. Ibrahim, R. Weerasuriya, D. Hillerkuss, G. Zarris, D. Simeonidou, J. Leuthold, D. Cotter, and A. D. Ellis, "Experimental demonstration of 42.6 Gbit/s asynchronous digital optical regenerators," in *International Conference on Transparent Optical Networks* (2008), paper We.C3.3.

[H6] J. Leuthold, W. Freude, S. Sygletos, P. Vorreau, R. Bonk, D. Hillerkuss, I. Tomkos, A. Tzanakaki, C. Kouloumentas, D. J. Richardson, P. Petropoulos, F. Parmigiani, A. D. Ellis, D. Cotter, S. Ibrahim, and R. Weerasuriya, "An all-optical grooming switch to interconnect access and metro ring networks," in *International Conference on Transparent Optical Networks* (2008), paper We.C3.4.

[H7] P. Vorreau, D. Hillerkuss, F. Parmigiani, S. Sygletos, R. Bonk, P. Petropoulos, D. Richardson, G. Zarris, D. Simeonidou, D. Klonidis, I. Tomkos, R. Weerasuriya, S. Ibrahim, A. Ellis, R. Morais, P. Monteiro, S. Ben Ezra, S. Tsadka, W. Freude, and J. Leuthold, "2R/3R optical grooming switch with time-slot interchange," in *European Conference on Optical Communication* (2008), paper PDP Th.3.F.4. .

[H8] J. Leuthold, R. Bonk, P. Vorreau, S. Sygletos, D. Hillerkuss, W. Freude, G. Zarris, D. Simeonidou, C. Kouloumentas, M. Spyropoulou, I. Tomkos, F. Parmigiani, P. Petropoulos, D. J. Richardson, R. Weerasuriya, S. Ibrahim, A. D. Ellis, C. Meuer, D. Bimberg, R. Morais, P. Monteiro, S. Ben-Ezra, and S. Tsadka, "An all-optical grooming switch with regenerative capabilities," in *International Conference on Transparent Optical Networks* (2009), paper We.A3.4.

[H9] J. Li, K. Worms, P. Vorreau, D. Hillerkuss, A. Ludwig, R. Maestle, S. Schule, U. Hollenbach, J. Mohr, W. Freude, and J. Leuthold, "Optical vector signal analyzer based on differential direct detection," in *LEOS Annual Meeting* (2009), paper TuA4.

[H10] A. Marculescu, S. Sygletos, J. Li, D. Karki, D. Hillerkuss, S. Ben-Ezra, S. Tsadka, W. Freude, and J. Leuthold, "RZ to CSRZ format and wavelength conversion with regenerative properties," in *Optical Fiber Communication Conference* (2009), paper OThS1.

[H11] T. Vallaitis, D. Hillerkuss, J. S. Li, R. Bonk, N. Lindenmann, P. Dumon, R. Baets, M. L. Scimeca, I. Biaggio, F. Diederich, C. Koos, W. Freude, and J. Leuthold, "All-optical wavelength conversion using cross-phase modulation at 42.7 Gbit/s in silicon-organic hybrid (SOH) waveguides," in *Photonics in Switching* (2009), pp. 1-2.

[H12] P. Vorreau, S. Sygletos, F. Parmigiani, D. Hillerkuss, R. Bonk, P. Petropoulos, D. J. Richardson, G. Zarris, D. Simeonidou, D. Klonidis, I. Tomkos, R. Weerasuriya, S. Ibrahim, A. D. Ellis, D. Cotter, R. Morais, P. Monteiro, S. Ben-Ezra, S. Tsadka, W. Freude, and J. Leuthold, "Optical grooming switch with regenerative functionality for transparent interconnection of networks," Opt. Express **17**(17), 15173-15185 (2009).

[H13] G. Zarris, F. Parmigiani, E. Hugues-Salas, R. Weerasuriya, D. Hillerkuss, N. A. Gonzalez, M. Spyropoulou, P. Vorreau, R. Morais, S. K. Ibrahim, D. Klonidis, P. Petropoulos, A. D. Ellis, P. Monteiro, A. Tzanakaki, D. Richardson, I. Tomkos, R. Bonk, W. Freude, J. Leuthold, and D. Simeonidou, "Field trial of WDM-OTDM transmultiplexing employing photonic switch fabric-based buffer-less bit-interleaved data grooming and all-optical regeneration," in *Optical Fiber Communication Conference* (2009), paper PDPC10.

[H14] G. Zarris, P. Vorreau, D. Hillerkuss, S. K. Ibrahim, R. Weerasuriya, A. D. Ellis, J. Leuthold, and D. Simeonidou, "WDM-to-OTDM traffic grooming by means of asynchronous retiming," in *Optical Fiber Communication Conference* (2009), paper OThJ6.

[H15] L. Alloatti, D. Korn, D. Hillerkuss, T. Vallaitis, J. Li, R. Bonk, R. Palmer, T. Schellinger, A. Barklund, R. Dinu, J. Wieland, M. Fournier, J. Fedeli, P. Dumon, R. Baets, C. Koos, W. Freude, and J. Leuthold, "40 Gbit/s silicon-organic hybrid (SOH) phase modulator," in *European Conference on Optical Communication* (2010), paper Tu.5.C.4.

[H16] L. Alloatti, D. Korn, D. Hillerkuss, T. Vallaitis, J. Li, R. Bonk, R. Palmer, T. Schellinger, C. Koos, W. Freude, J. Leuthold, M. Fournier, J. Fedeli, A. Barklund, R. Dinu, J. Wieland, W. Bogaerts, P. Dumon, and R. Baets, "Silicon high-speed electro-optic modulator," in *IEEE International Conference on Group IV Photonics* (2010), pp. 195-197.

[H17] R. Bonk, T. Vallaitis, J. Guetlein, D. Hillerkuss, J. Li, W. Freude, and J. Leuthold, "Quantum dot SOA dynamic range improvement for phase modulated signals," in *Optical Fiber Communication Conference* (2010), paper OThK3.

[H18] W. Freude, L. Alloatti, T. Vallaitis, D. Korn, D. Hillerkuss, R. Bonk, R. Palmer, J. Li, T. Schellinger, M. Fournier, J. Fedeli, W. Bogaerts, P. Dumon, R. Baets, A. Barklund, R. Dinu, J. Wieland, M. L. Scimeca, I. Biaggio, B. Breiten, F. Diederich, C. Koss, and J. Leuthold, "High-speed signal processing with silicon-organic hybrid devices," in *Journal of the European Optical Society: rapid publications* (2010).

[H19] R. Freund, M. Nölle, C. Schmidt-Langhorst, R. Ludwig, C. Schubert, G. Bosco, A. Carena, P. Poggiolini, L. Oxenløwe, M. Galili, H. C. H. Mulvad, M. Winter, D. Hillerkuss, R. Schmogrow, W. Freude, J. Leuthold, A. D. Ellis, F. C. G. Gunning, J. Zhao, P. Frascella, S. K. Ibrahim, and N. M. Suibhne, "Single- and multi-carrier techniques to build up Tb/s per channel transmission systems," in *International Conference on Transparent Optical Networks* (2010), paper TU.D1.4.

[H20] D. Hillerkuss, A. Marculescu, J. Li, M. Teschke, G. Sigurdsson, K. Worms, S. Ben-Ezra, N. Narkiss, W. Freude, and J. Leuthold, "Novel optical fast Fourier transform scheme enabling real-time OFDM processing at 392 Gbit/s and beyond," in *Optical Fiber Communication Conference* (2010), paper OWW3.

[H21] D. Hillerkuss, T. Schellinger, R. Schmogrow, M. Winter, T. Vallaitis, R. Bonk, A. Marculescu, J. Li, M. Dreschmann, J. Meyer, S. Ben-Ezra, N. Narkiss, B. Nebendahl, F. Parmigiani, P. Petropoulos, B. Resan, K. Weingarten, T. Ellermeyer, J. Lutz, M. Moller, M. Huebner, J. Becker, C. Koos, W. Freude, and J. Leuthold, "Single source optical OFDM transmitter and optical FFT receiver demonstrated at line rates of 5.4 and 10.8 Tbit/s," in *Optical Fiber Communication Conference* (2010), paper PDPC1.

[H22] D. Hillerkuss, R. Schmogrow, M. Huebner, M. Winter, B. Nebendahl, J. Becker, W. Freude, and J. Leuthold, "Software-defined multi-format transmitter with real-time signal processing for up to 160 Gbit/s," in *Signal Processing in Photonic Communications* (Optical Society of America, 2010), paper SPTuC4.

[H23] D. Hillerkuss, M. Winter, M. Teschke, A. Marculescu, J. Li, G. Sigurdsson, K. Worms, S. Ben Ezra, N. Narkiss, W. Freude, and J. Leuthold, "Simple all-optical FFT scheme enabling Tbit/s real-time signal processing," Opt. Express **18**(9), 9324-9340 (2010).

[H24] D. Hillerkuss, M. Winter, M. Teschke, A. Marculescu, J. Li, G. Sigurdsson, K. Worms, W. Freude, and J. Leuthold, "Low-complexity optical FFT scheme enabling Tbit/s all-optical OFDM communication," ITG Symposium on Photonic Networks, 1-8 (2010).

[H25] J. Leuthold, R. Bonk, P. Vorreau, S. Sygletos, D. Hillerkuss, W. Freude, G. Zarris, D. Simeonidou, C. Kouloumentas, M. Spyropoulou, I. Tomkos, F. Parmigiani, P. Petropoulos, D. J. Richardson, R. Weerasuriya, S. Ibrahim, A. D. Ellis, R. Morais, P. Monteiro, S. B. Ezra, and S. Tsadka, "All-optical grooming for 100 Gbit/s ethernet," in *P Soc Photo-Opt Ins*, W. Weiershausen, B. Dingel, A. K. Dutta, and A. K. Srivastava, eds. (SPIE, 2010), pp. 762107-762108.

[H26] J. Leuthold, D. Hillerkuss, M. Winter, J. Li, K. Worms, C. Koos, W. Freude, S. Ben-Ezra, and N. Narkiss, "Terabit/s FFT processing – optics can do it on-the-fly," in *International Conference on Transparent Optical Networks* (2010), paper Mo.D1.4.

[H27] J. Leuthold, M. Winter, W. Freude, C. Koos, D. Hillerkuss, T. Schellinger, R. Schmogrow, T. Vallaitis, R. Bonk, A. Marculescu, J. Li, M. Dreschmann, J. Meyer, M. Huebner, J. Becker, S. Ben-Ezra, N. Narkiss, B. Nebendahl, F. Parmigiani, P. Petropoulos, B. Resan, A. Oehler, K. Weingarten, T. Ellermeyer, J. Lutz, and M. Möller, "All-optical FTT signal processing of a 10.8 Tb/s single channel OFDM signal," in *Photonics in Switching* (Optical Society of America, 2010), paper PWC1.

[H28] J. Li, K. Worms, D. Hillerkuss, B. Richter, R. Maestle, W. Freude, and J. Leuthold, "Tunable free space optical delay interferometer for demodulation of differential phase shift keying signals," in *Optical Fiber Communication Conference* (2010), paper JWA24.

[H29] J. Li, K. Worms, A. Marculescu, D. Hillerkuss, S. Ben-Ezra, W. Freude, and J. Leuthold, "Optical vector signal analyzer based on differential detection with inphase and quadrature phase control," in *Signal Processing in Photonic Communications* (Optical Society of America, 2010), paper SPTuB3.

[H30] R. Schmogrow, D. Hillerkuss, M. Dreschmann, M. Huebner, M. Winter, J. Meyer, B. Nebendahl, C. Koos, J. Becker, W. Freude, and J. Leuthold, "Real-time software-defined multiformat transmitter generating 64QAM at 28 GBd," IEEE Photon. Technol. Lett. **22**(21), 1601-1603 (2010).

[H31] M. Spyropoulou, R. Bonk, D. Hillerkuss, N. Pleros, T. Vallaitis, W. Freude, I. Tomkos, and J. Leuthold, "Experimental investigation of multi-wavelength clock recovery based on a quantum-dot SOA at 40 Gb/s," in *Signal Processing in Photonic Communications* (Optical Society of America, 2010), paper SPTuB4.

[H32] T. Vallaitis, R. Bonk, J. Guetlein, D. Hillerkuss, J. Li, R. Brenot, F. Lelarge, G. H. Duan, W. Freude, and J. Leuthold, "Quantum dot SOA input power dynamic range improvement for differential-phase encoded signals," Opt. Express **18**(6), 6270-6276 (2010).

[H33] T. Vallaitis, R. Bonk, J. Guetlein, D. Hillerkuss, J. Li, W. Freude, J. Leuthold, C. Koos, M. L. Scimeca, I. Biaggio, F. Diederich, B. Breiten, P. Dumon, and R. Baets,

"All-optical wavelength conversion of 56 Gbit/s NRZ-DQPSK signals in silicon-organic hybrid strip waveguides," in *Optical Fiber Communication Conference* (2010), paper OTuN1.

[H34] T. Vallaitis, R. Bonk, J. Guetlein, C. Meuer, D. Hillerkuss, W. Freude, D. Bimberg, and J. Leuthold, "Optimizing SOA for large input power dynamic range with respect to applications in extended GPON," in *Access Networks and In-house Communications* (Optical Society of America, 2010), paper AThC4.

[H35] G. Zarris, E. Hugues-Salas, N. A. Gonzalez, R. Weerasuriya, F. Parmigiani, D. Hillerkuss, P. Vorreau, M. Spyropoulou, S. K. Ibrahim, A. D. Ellis, R. Morais, P. Monteiro, P. Petropoulos, D. J. Richardson, I. Tomkos, J. Leuthold, and D. Simeonidou, "Field experiments with a grooming switch for OTDM meshed networking," J. Lightwave Technol. **28**(4), 316-327 (2010).

[H36] L. Alloatti, D. Korn, R. Palmer, D. Hillerkuss, J. Li, A. Barklund, R. Dinu, J. Wieland, M. Fournier, J. Fedeli, H. Yu, W. Bogaerts, P. Dumon, R. Baets, C. Koos, W. Freude, and J. Leuthold, "42.7 Gbit/s electro-optic modulator in silicon technology," Opt. Express **19**(12), 11841-11851 (2011).

[H37] R. Bonk, G. Huber, T. Vallaitis, R. Schmogrow, D. Hillerkuss, C. Koos, W. Freude, and J. Leuthold, "Impact of alfa-factor on SOA dynamic range for 20 GBd BPSK, QPSK and 16-QAM signals," in *Optical Fiber Communication Conference* (Optical Society of America, 2011), paper OML4.

[H38] R. Bonk, P. Vorreau, D. Hillerkuss, W. Freude, G. Zarris, D. Simeonidou, F. Parmigiani, P. Petropoulos, R. Weerasuriya, S. Ibrahim, A. D. Ellis, D. Klonidis, I. Tomkos, and J. Leuthold, "An all-optical grooming switch for interconnecting access and metro ring networks [Invited]," J. Opt. Commun. Netw. **3**(3), 206-214 (2011).

[H39] M. Dreschmann, J. Meyer, M. Hubner, R. Schmogrow, D. Hillerkuss, J. Becker, J. Leuthold, and W. Freude, "Implementation of an ultra-high speed 256-point FFT for Xilinx Virtex-6 devices," in *IEEE International Conference on Industrial Informatics (INDIN)* (2011), pp. 829-834.

[H40] W. Freude, D. Hillerkuss, T. Schellinger, R. Schmogrow, M. Winter, T. Vallaitis, R. Bonk, A. Marculescu, J. Li, M. Dreschmann, J. Meyer, S. Ben-Ezra, M. Caspi, B. Nebendahl, F. Parmigiani, P. Petropoulos, B. Resan, A. E. H. Oehler, K. Weingarten, T. Ellermeyer, J. Lutz, M. Moeller, M. Huebner, J. Becker, C. Koos, and J. Leuthold, "All-optical real-time OFDM transmitter and receiver," in *Conference on Lasers and Electro-Optics (CLEO)* (Optical Society of America, 2011), paper CThO1.

[H41] W. Freude, R. Schmogrow, B. Nebendahl, D. Hillerkuss, J. Meyer, M. Dreschmann, M. Huebner, J. Becker, C. Koos, and J. Leuthold, "Software-defined optical transmission," in *International Conference on Transparent Optical Networks* (2011), paper TU.D1.1.

[H42] D. Hillerkuss, R. Schmogrow, T. Schellinger, M. Jordan, M. Winter, G. Huber, T. Vallaitis, R. Bonk, P. Kleinow, F. Frey, M. Roeger, S. Koenig, A. Ludwig, A. Marculescu, J. Li, M. Hoh, M. Dreschmann, J. Meyer, S. Ben-Ezra, N. Narkiss, B. Nebendahl, F. Parmigiani, P. Petropoulos, B. Resan, A. Oehler, K. Weingarten, T. Ellermeyer, J. Lutz, M. Moeller, M. Huebner, J. Becker, C. Koos, W. Freude, and J. Leuthold, "26 Tbit s^{-1} line-rate super-channel transmission utilizing all-optical fast Fourier transform processing," Nature Photon. **5**(6), 364-371 (2011).

[H43] C. Koos, L. Alloatti, D. Korn, R. Palmer, D. Hillerkuss, J. Li, A. Barklund, R. Dinu, J. Wieland, M. Fournier, J.-M. Fedeli, H. Yu, W. Bogaerts, P. Dumon, R. Baets, W. Freude, and J. Leuthold, "Silicon-organic hybrid (SOH) electro-optical devices," in *Integrated Photonics Research, Silicon and Nanophotonics* (Optical Society of America, 2011), paper IWF1.

[H44] C. Koos, L. Alloatti, D. Korn, R. Palmer, T. Vallaitis, R. Bonk, D. Hillerkuss, J. Li, W. Bogaerts, P. Dumon, R. Baets, M. L. Scimeca, I. Biaggio, A. Barklund, R. Dinu, J. Wieland, M. Fournier, J. Fedeli, W. Freude, and J. Leuthold, "Silicon nanophotonics and silicon-organic hybrid (SOH) integration," in *General Assembly and Scientific Symposium (URSI)* (2011), pp. 1-4.

[H45] J. Leuthold, W. Freude, C. Koos, R. Bonk, S. Koenig, D. Hillerkuss, and R. Schmogrow, "Semiconductor optical amplifiers in extended reach PONs," (Optical Society of America, 2011), paper ATuA1.

[H46] J. Leuthold, W. Freude, C. Koos, D. Hillerkuss, R. Schmogrow, and S. Ben-Ezra, "Terabit/s super-channels based on OFDM," in *Signal Processing in Photonic Communications* (Optical Society of America, 2011), paper SPMB1.

[H47] J. Li, C. Schmidt-Langhorst, R. Schmogrow, D. Hillerkuss, M. Lauermann, M. Winter, K. Worms, C. Schubert, C. Koos, W. Freude, and J. Leuthold, "Self-coherent receiver for PolMUX coherent signals," in *Optical Fiber Communication Conference* (Optical Society of America, 2011), paper OWV5.

[H48] J. Li, K. Worms, R. Maestle, D. Hillerkuss, W. Freude, and J. Leuthold, "Free-space optical delay interferometer with tunable delay and phase," Opt. Express **19**(12), 11654-11666 (2011).

[H49] N. Lindenmann, G. Balthasar, D. Hillerkuss, R. Schmogrow, M. Jordan, J. Leuthold, W. Freude, and C. Koos, "Photonic wire bonds for terabit/s chip-to-chip interconnects," ArXiv e-prints, arXiv:1111.0651v1 (2011),

http://arxiv.org/abs/1111.0651v1.

[H50] N. Lindenmann, I. Kaiser, G. Balthasar, R. Bonk, D. Hillerkuss, W. Freude, J. Leuthold, and C. Koos, "Photonic waveguide bonds - A novel concept for chip-to-chip interconnects," in *Optical Fiber Communication Conference* (2011), pp. 1-3.

[H51] R. Schmogrow, M. Winter, D. Hillerkuss, B. Nebendahl, S. Ben-Ezra, J. Meyer, M. Dreschmann, M. Huebner, J. Becker, C. Koos, W. Freude, and J. Leuthold, "Real-time OFDM transmitter beyond 100 Gbit/s," Opt. Express **19**(13), 12740-12749 (2011).

[H52] R. Schmogrow, M. Winter, M. Meyer, D. Hillerkuss, B. Nebendahl, J. Meyer, M. Dreschmann, M. Huebner, J. Becker, C. Koos, W. Freude, and J. Leuthold, "Real-time Nyquist pulse modulation transmitter generating rectangular shaped spectra of 112 Gbit/s 16QAM signals," in *Signal Processing in Photonic Communications* (Optical Society of America, 2011), paper SPMA5.

[H53] R. Schmogrow, M. Winter, B. Nebendahl, D. Hillerkuss, J. Meyer, M. Dreschmann, M. Huebner, J. Becker, C. Koos, W. Freude, and J. Leuthold, "101.5 Gbit/s real-time OFDM transmitter with 16QAM modulated subcarriers," in *Optical Fiber Communication Conference* (2011), paper OWE5.

[H54] R. Bonk, G. Huber, T. Vallaitis, S. Koenig, R. Schmogrow, D. Hillerkuss, R. Brenot, F. Lelarge, G. H. Duan, S. Sygletos, C. Koos, W. Freude, and J. Leuthold, "Linear semiconductor optical amplifiers for amplification of advanced modulation formats," Opt. Express **20**(9), 9657-9672 (2012).

[H55] R. Bouziane, R. Schmogrow, D. Hillerkuss, P. Milder, C. Koos, W. Freude, J. Leuthold, P. Bayvel, and R. Killey, "Generation and transmission of 85.4 Gb/s real-time 16QAM coherent optical OFDM signals over 400 km SSMF with preamble-less reception," Opt. Express **20**(19), 21612-21617 (2012).

[H56] M. Dreschmann, J. Meyer, M. Hubner, R. Schmogrow, D. Hillerkuss, J. Becker, J. Leuthold, and W. Freude, "Time and frequency synchronization for ultra-high speed

OFDM systems," in *International Conference on Computing, Networking and Communications (ICNC)* (2012), pp. 871-875.

[H57] W. Freude, L. Alloatti, A. Melikyan, R. Palmer, D. Korn, N. Lindenmann, T. Vallaitis, D. Hillerkuss, J. Li, A. Barklund, R. Dinu, J. Wieland, M. Fournier, J.-M. Fedeli, S. Walheim, P. Leufke, S. Ulrich, J. Ye, P. Vincze, H. Hahn, H. Yu, W. Bogaerts, P. Dumon, R. Baets, B. Breiten, F. Diederich, M. T. Beels, I. Biaggio, T. Schimmel, C. Koos, and J. Leuthold, "Nonlinear optics on the silicon platform," in *Optical Fiber Communication Conference* (Optical Society of America, 2012), paper OTh3H.6.

[H58] W. Freude, R. Schmogrow, D. Hillerkuss, J. Meyer, M. Dreschmann, B. Nebendahl, M. Huebner, J. Becker, C. Koos, and J. Leuthold, "Reconfigurable optical transmitters and receivers," in *Photonics West*, G. Li, and D. S. Jager, eds. (SPIE, 2012), pp. 82840A-82848.

[H59] W. Freude, R. Schmogrow, B. Nebendahl, M. Winter, A. Josten, D. Hillerkuss, S. Koenig, J. Meyer, M. Dreschmann, M. Huebner, C. Koos, J. Becker, and J. Leuthold, "Quality metrics for optical signals: eye diagram, Q-factor, OSNR, EVM and BER," in *International Conference on Transparent Optical Networks* (2012), paper Mo.B1.5.

[H60] D. Hillerkuss, R. Schmogrow, C. Koos, W. Freude, J. Leuthold, B. Nebendahl, and T. Dippon, "Real-time Nyquist Pulse Transmission," ITG Symposium on Photonic Networks (2012).

[H61] D. Hillerkuss, R. Schmogrow, M. Meyer, S. Wolf, M. Jordan, P. Kleinow, N. Lindenmann, P. Schindler, A. Melikyan, X. Yang, S. Ben-Ezra, B. Nebendahl, M. Dreschmann, J. Meyer, F. Parmigiani, P. Petropoulos, B. Resan, A. Oehler, K. Weingarten, L. Altenhain, T. Ellermeyer, M. Moeller, M. Huebner, J. Becker, C. Koos, W. Freude, and J. Leuthold, "Single-laser 32.5 Tbit/s Nyquist WDM transmission," ArXiv e-prints arXiv:1203.2516v1 (2012),

http://arxiv.org/abs/1203.2516v1.

[H62] D. Hillerkuss, R. Schmogrow, M. Meyer, S. Wolf, M. Jordan, P. Kleinow, N. Lindenmann, P. Schindler, A. Melikyan, X. Yang, S. Ben-Ezra, B. Nebendahl, M. Dreschmann, J. Meyer, F. Parmigiani, P. Petropoulos, B. Resan, A. Oehler, K. Weingarten, L. Altenhain, T. Ellermeyer, M. Moeller, M. Huebner, J. Becker, C. Koos, W. Freude, and J. Leuthold, "Single-laser 32.5 Tbit/s Nyquist WDM transmission," J. Opt. Commun. Netw. **4**(10), 715-723 (2012).

[H63] D. Hillerkuss, R. Schmogrow, M. Meyer, S. Wolf, M. Jordan, P. Kleinow, N. Lindenmann, P. C. Schindler, A. Melikyan, M. Dreschmann, J. Meyer, J. Becker, C. Koos, W. Freude, J. Leuthold, X. Yang, F. Parmigiani, P. Petropoulos, S. Ben-Ezra, B. Nebendahl, B. Resan, A. Oehler, K. Weingarten, L. Altenhain, T. Ellermeyer, M. Moeller, and M. Huebner, "Single-laser 32.5 Tbit/s Nyquist-WDM transmission over 227 km with real-time Nyquist pulse shaping," in *Photonics in Switching* (2012), pp. Th-S13-I06.

[H64] Y. Hui, M. Pantouvaki, J. Van Campenhout, K. Komorowska, P. Dumon, P. Verheyen, G. Lepage, P. Absil, D. Korn, D. Hillerkuss, J. Leuthold, R. Baets, and W. Bogaerts, "Silicon carrier-depletion-based Mach-Zehnder and ring modulators with different doping patterns for telecommunication and optical interconnect," in *International Conference on Transparent Optical Networks* (2012), pp. 1-5.

[H65] D. Korn, H. Yu, D. Hillerkuss, L. Alloatti, C. Mattern, W. Bogaerts, K. Komorowska, R. Baets, J. Van Campenhout, P. Verheyen, J. Wouters, M. Moelants, P. Absil, W. Freude, C. Koos, and J. Leuthold, "Detection or modulation at 35 Gbit/s with a standard CMOS-processed optical waveguide," in *Conference on Lasers and Electro-Optics (CLEO)* (Optical Society of America, 2012), paper CTu1A.1.

[H66] J. Leuthold, R. Schmogrow, D. Hillerkuss, C. Koos, and W. Freude, "Super channels based on Nyquist multiplexing," in *Opto-Electronics and Communications Conference (OECC)* (2012), pp. 30-32.

[H67] J. Li, R. Schmogrow, D. Hillerkuss, P. C. Schindler, M. Nazarthy, C. Schmidt-Langhorst, S.-B. Ezra, I. Tselniker, C. Koos, and W. Freude, "A self-coherent receiver for detection of PolMUX coherent signals," Opt. Express **20**(19), 21413-21433 (2012).

[H68] N. Lindenmann, G. Balthasar, D. Hillerkuss, R. Schmogrow, M. Jordan, J. Leuthold, W. Freude, and C. Koos, "Photonic wire bonding: a novel concept for chip-scale interconnects," Opt. Express **20**(16), 17667-17677 (2012).

[H69] N. Lindenmann, G. Balthasar, M. Jordan, D. Hillerkuss, R. Schmogrow, W. Freude, J. Leuthold, and C. Koos, "Low-loss photonic wire bond interconnects enabling 5 TBit/s data transmission," in *Optical Fiber Communication Conference* (Optical Society of America, 2012), paper OW4I.4.

[H70] J. Meyer, S. Menzel, M. Dreschmann, R. Schmogrow, D. Hillerkuss, W. Freude, J. Leuthold, and J. Becker, "Ultra High Speed Digital Down Converter Design for Virtex-6 FPGAs," in *International OFDM Workshop (InOWo)* (VDE, 2012), pp. 1-5.

[H71] A. K. Mishra, Z. Wang, H. Klein, R. Bonk, S. Koenig, D. Karnick, R. Schmogrow, D. Hillerkuss, M. Moehrle, T. Pfeiffer, C. Koos, J. Leuthold, and W. Freude, "Performance analysis of an OFDM transmission system with directly modulated lasers for wireless backhauling," in *International Conference on Transparent Optical Networks* (2012), pp. 1-4.

[H72] B. Nebendahl, R. Schmogrow, T. Dennis, A. Josten, D. Hillerkuss, S. Koenig, J. Meyer, M. Dreschmann, M. Winter, and M. Huebener, "Quality Metrics in Optical Modulation Analysis: EVM and its relation to Q-factor, OSNR, and BER," in *Asia Communications and Photonics Conference* (Optical Society of America, 2012).

[H73] J. Pfeifle, C. Weimann, F. Bach, J. Riemensberger, K. Hartinger, D. Hillerkuss, M. Jordan, R. Holtzwarth, T. J. Kippenberg, J. Leuthold, W. Freude, and C. Koos, "Microresonator-based optical frequency combs for high-bitrate WDM data transmission," in *Optical Fiber Communication Conference* (Optical Society of America, 2012), paper OW1C.4.

[H74] P. C. Schindler, R. M. Schmogrow, D. Hillerkuss, M. Nazarathy, S. Ben-Ezra, C. Koos, W. Freude, and J. Leuthold, "Remote Heterodyne Reception of OFDM-QPSK as Downlink-Solution for Future Access Networks," in *Access Networks and In-house Communications (ANIC)* (Optical Society of America, 2012), paper AW4A.3.

[H75] R. Schmogrow, R. Bouziane, D. Hillerkuss, P. Milder, R. J. Koutsoyannis, R. Killey, Y. Benlachtar, P. Watts, P. Bayvel, C. Koos, W. Freude, and J. Leuthold, "85.4 Gbit/s real-time OFDM signal generation with transmission over 400 km and preamble-less reception," in *Conference on Lasers and Electro-Optics (CLEO)* (Optical Society of America, 2012), paper CTh1H.4.

[H76] R. Schmogrow, D. Hillerkuss, C. Koos, W. Freude, J. Leuthold, and B. Nebendahl, "Quality metrics for advanced modulation formats in optical communications: OSNR, Q-factor, EVM, and BER," ITG Symposium on Photonic Networks (2012).

[H77] R. Schmogrow, D. Hillerkuss, S. Wolf, B. Bäuerle, M. Winter, P. Kleinow, B. Nebendahl, T. Dippon, P. C. Schindler, C. Koos, W. Freude, and J. Leuthold, "512QAM Nyquist sinc-pulse transmission at 54 Gbit/s in an optical bandwidth of 3 GHz," Opt. Express **20**(6), 6439-6447 (2012).

[H78] R. Schmogrow, M. Meyer, S. Wolf, B. Nebendahl, D. Hillerkuss, B. Baeuerle, M. Dreschmann, J. Meyer, M. Huebner, J. Becker, C. Koos, W. Freude, and J. Leuthold,

"150 Gbit/s real-time Nyquist pulse transmission over 150 km SSMF enhanced by DSP with dynamic precision," in *Optical Fiber Communication Conference* (2012), paper OM2A.6.

[H79] R. Schmogrow, B. Nebendahl, M. Winter, A. Josten, D. Hillerkuss, S. Koenig, J. Meyer, M. Dreschmann, M. Huebner, and C. Koos, "Corrections to "Error Vector Magnitude as a Performance Measure for Advanced Modulation Formats"," IEEE Photon. Technol. Lett. **24**(23), 2198-2198 (2012).

[H80] R. Schmogrow, B. Nebendahl, M. Winter, A. Josten, D. Hillerkuss, S. Koenig, J. Meyer, M. Dreschmann, M. Huebner, C. Koos, J. Becker, W. Freude, and J. Leuthold, "Error vector magnitude as a performance measure for advanced modulation formats," IEEE Photon. Technol. Lett. **24**(1), 61-63 (2012).

[H81] R. Schmogrow, M. Winter, M. Meyer, A. Ludwig, D. Hillerkuss, B. Nebendahl, S. Ben-Ezra, J. Meyer, M. Dreschmann, M. Huebner, J. Becker, C. Koos, W. Freude, and J. Leuthold, "Real-time Nyquist pulse generation beyond 100 Gbit/s and its relation to OFDM," Opt. Express **20**(1), 317-337 (2012).

[H82] R. Schmogrow, S. Wolf, B. Baeuerle, D. Hillerkuss, B. Nebendahl, C. Koos, W. Freude, and J. Leuthold, "Nyquist frequency division multiplexing for optical communications," in *Conference on Lasers and Electro-Optics (CLEO)* (Optical Society of America, 2012), paper CTh1H.2.

[H83] R. M. Schmogrow, B. Baeuerle, D. Hillerkuss, B. Nebendahl, C. Koos, W. Freude, and J. Leuthold, "Raised-cosine OFDM for enhanced out-of-band suppression at low subcarrier counts," in *Signal Processing in Photonic Communications (SPPCOM)* (Optical Society of America, 2012), paper SpTu2A.2.

[H84] R. M. Schmogrow, P. C. Schindler, D. Hillerkuss, C. Koos, W. Freude, and J. Leuthold, "Uplink solutions for future access networks," in *Access Networks and In-house Communications (ANIC)* (Optical Society of America, 2012), paper AW4A.1.

[H85] H. Yu, W. Bogaerts, K. Komorowska, R. Baets, D. Korn, L. Alloatti, D. Hillerkuss, C. Koos, W. Freude, J. Leuthold, J. Campenhout, P. Verheyen, J. Wouters, M. Moelants, and P. Absil, "Doping geometries for 40G carrier-depletion-based Silicon optical modulators," in *Optical Fiber Communication Conference* (Optical Society of America, 2012), paper OW4F.4.

[H86] H. Yu, D. Korn, M. Pantouvaki, J. Van Campenhout, K. Komorowska, P. Verheyen, G. Lepage, P. Absil, D. Hillerkuss, and L. Alloatti, "Using carrier-depletion silicon modulators for optical power monitoring," Opt. Lett. **37**(22), 4681-4683 (2012).

[H87] H. Yu, M. Pantouvaki, J. Van Campenhout, D. Korn, K. Komorowska, P. Dumon, Y. Li, P. Verheyen, P. Absil, L. Alloatti, D. Hillerkuss, J. Leuthold, R. Baets, and W. Bogaerts, "Performance tradeoff between lateral and interdigitated doping patterns for high speed carrier-depletion based silicon modulators," Opt. Express **20**(12), 12926-12938 (2012).

[H88] D. Hillerkuss, T. Schellinger, M. Jordan, C. Weimann, F. Parmigiani, B. Resan, K. Weingarten, S. Ben-Ezra, B. Nebendahl, C. Koos, W. Freude, and J. Leuthold, "High-Quality Optical Frequency Comb by Spectral Slicing of Spectra Broadened by SPM," IEEE Photonics Journal **5**(5), 7201011-7201011 (2013).

Karlsruhe Series in Photonics & Communications
KIT, Institute of Photonics and Quantum Electronics (IPQ)
(ISSN 1865-1100)

Die Bände sind unter www.ksp.kit.edu als PDF frei verfügbar
oder als Druckausgabe bestellbar.

Karlsruhe Series in Photonics & Communications
KIT, Institute of Photonics and Quantum Electronics (IPQ)
(ISSN 1865-1100)

Band 9   David Hillerkuss
Single-Laser Multi-Terabit/s Systems. 2013
ISBN 978-3-86644-991-6